T0297786

GENERA OF BRITISH PLANTS

GENERA OF BRITISH PLANTS

ARRANGED ACCORDING TO

ENGLER'S *SYLLABUS DER PFLANZENFAMILIEN*

(SEVENTH EDITION 1912)

WITH THE ADDITION OF THE CHARACTERS OF THE GENERA

BY

HUMPHREY G. CARTER, M.B., Ch.B.

Cambridge:
at the University Press
1913

CAMBRIDGE
UNIVERSITY PRESS

University Printing House, Cambridge CB2 8BS, United Kingdom

Cambridge University Press is part of the University of Cambridge.

It furthers the University's mission by disseminating knowledge in the pursuit of education, learning and research at the highest international levels of excellence.

www.cambridge.org
Information on this title: www.cambridge.org/9781316509753

© Cambridge University Press 1913

First published 1913
First paperback edition 2015

A catalogue record for this publication is available from the British Library

ISBN 978-1-316-50975-3 Paperback

PREFACE

WHAT is known as Engler's System of the Classification of Plants embodied in the *Natürliche Pflanzenfamilien* and in the successive editions of the *Syllabus der Pflanzenfamilien*, in the opinion of many modern systematic botanists represents, as a whole, the nearest approach to a natural system that we possess.

While specialists may criticise it in one detail or another, and while it is certain that we are as yet far from finality, owing to the numerous lacunae which still exist in our knowledge of different groups, there can be little if any doubt that no other published system of classification rivals Professor Engler's in broad grasp of principle and masterly treatment of complicated detail. It is the conspicuous merit of Professor Engler and his collaborators that they have kept constantly presented to botanists in the successive editions of the *Syllabus* the best scheme that could in their judgment be devised at the moment.

The present work is intended to familiarise students of British Vascular Plants with Engler's system in its latest form and thus both to habituate British floristic students to the use of a more natural system than that to which they have been accustomed in the British Floras that have hitherto appeared, and to serve as an introduction to the use of the *Syllabus*, the *Natürliche Pflanzenfamilien* and the *Pflanzenreich.*

The major divisions of the vascular plants with their diagnoses are taken, to a large extent, *verbatim* from the *Syllabus der Pflanzenfamilien* (7th Edition 1912).

Professor Engler and his publishers have most kindly and readily accorded their assent to this very free use of their publication.

In a few cases Engler's arrangement has been departed from.

Thus the nomenclature of the larger divisions of Pteridophyta has been somewhat modified in order to secure uniformity in the terminations of the names of Orders (Reihen). Prof. Bower's arrangement of the Leptosporangiate Ferns into Simplices Mixtae and Gradatae has been adopted as representing the most rational grouping, though the distinct suborder of Filicales, Hydropteridineae, has been retained owing to the difficulty of assigning with certainty the Marsiliaceae and Salviniaceae to their proper places in the Eufilianean series.

The Family Polypodiaceae, practically equivalent to the Mixtae of Bower, has been retained and divided into tribes according to the arrangement of the *Syllabus*, since our knowledge of the Systematics of this intricate group is not sufficiently advanced to admit of a grouping of the Mixtae into Natural Families. In dealing with the British Genera alone this arrangement does not appear to do much violence to natural affinities so far as they have been ascertained.

Among the Flowering Plants Warming's arrangement of the Urticales has been followed mainly because of the difficulty of defining the Family Moraceae so as to include Cannabis and Humulus (subfamily Cannaboideae of Engler).

In all the other Orders the arrangement of the *Syllabus* has been closely followed, though the actual diagnoses, especially in cases where an Order or Family is poorly represented in Britain, have been to some extent modified.

The Characters of the Genera have either been obtained from various sources, such as Hooker's *Student's Flora* and Garcke's *Flora von Deutschland*, or are original, and they will, it is hoped, prove useful to students.

The following books, besides the *Syllabus* and *Pflanzenfamilien*, have been constantly used by me during the compilation of this work.

Ascherson and Graebner. *Flora des Nordostdeutschen Flachlandes*, 1898–9.

Blytt. *Haandbog i Norges Flora*, 1906.

Garcke. *Flora von Deutschland*. Edited by Dr Franz Niedenzu, 1908.

Hooker, Sir J. D. *The Student's Flora of the British Islands*, 1884.

Rendle. *The Classification of Flowering Plants.* Vol. I.

Warming. *Frøplanterne*, 1912.

Willis, J. C. *Flowering Plants and Ferns*, 1908.

My best thanks are due to Mr Tansley for help and advice with the Pteridophytes, to Professor Seward for reading the proofs of the Pteridophytes and Gymnosperms, and to Dr Moss who has read the entire proofs and given me much valuable advice and assistance throughout.

H. G. C.

CAMBRIDGE,

    August 1913.

SYNOPTIC INDEX

The numbers in the first column refer to pages in this book, those in the second column refer to pages in Engler's Syllabus (Ed. 7, 1912).

GLOSSARY OF WORDS RELATING TO PERIANTH

Achlamydeous.	With no perianth (naked).
Apochlamydeous.	Perianth absent by reduction.
Apopetalous.	Corolla absent by reduction.
Haplochlamydeous.	Having a perianth consisting of a single whorl.
Diplochlamydeous.	Having a perianth consisting of two whorls.
Homochlamydeous.	Diplochlamydeous with both whorls of the same kind.
Heterochlamydeous.	Diplochlamydeous with whorls clearly unlike each other.

In Floral Formulae

K = Calyx.   C = Corolla.   A = Androecium.   G = Gynaecium.

* = not native.

⊙ = doubtfully native.

Division 1. **EMBRYOPHYTA ASIPHONO-GAMA (ARCHEGONIATAE).**

Plants with marked alternation of generations. *Sexual generation* (*gametophyte*) bearing *antheridia*, in which *spermatozoids* arise, and *archegonia*, each containing an *egg-cell*. The fertilized egg-cell gives rise to an *embryo*, which for some time or throughout life remains attached to, and nutritively dependent on, the gametophyte. The embryo grows into the asexual generation (*sporophyte*), eventually producing *spores*, which again give rise to the gametophytes.

[Subdivision 1. **Bryophyta (not included in this work).** Gametophyte thalloid or leafy, free living. Sporophyte (sporogonium) leafless, attached throughout life to the gametophyte : no true roots.]

Subdivision 2. **PTERIDOPHYTA.** *Gametophyte* (*prothallus*) thalloid, usually vegetating independently, rarely completely enclosed in the spore. *Sporophyte* a cormophytic plant with true leaves and roots, early becoming independent of the gametophyte, and bearing *spores* in separate organs (*sporangia*) borne on or in connection with the leaves.

Class 1. **EQUISETARIAE.** *Leaves* generally small in relation to the stem, in alternating whorls. *Stele* with separate collateral bundles. Peltate *sporangiophores* aggregated in cones. *Sporangia* bulky, attached to the

c.   1

2   *Equisetaceae, Lycopodiaceae*

inner surface of the lamina of the sporangiophore. Homosporous or (in some extinct forms) heterosporous. Order **Equisetales** (including all living forms). Homosporous.

Family EQUISETACEAE. *Rhizomes* subterranean often bearing distinct sterile and fertile shoots. Bases of leaf-whorls united into sheaths enclosing the base of the internode above. *Branching* axillary. *Epispore* splitting into two spirally wound threads (*elaters*). *Prothalli* branched, often functionally dioecious. *Antheridia* and *archegonia* sunken. *Spermatozoids* multiciliate.

Only genus: **Equisetum.**

Class 2. LYCOPODIARIAE. *Leaves* usually small in relation to stem. *Stem* with one or more haplosteles†. *Sporangia* single in axils, or on adaxial faces of fertile leaves. Homosporous or heterosporous. *Spermatozoids* biciliate.

Order 1. **Lycopodiales.** Small leaved herbs of erect or creeping habit. *Sporophylls* mostly aggregated in *cones*.

Suborder 1. ELIGULATAE. Homosporous forms without *ligule*.

Family LYCOPODIACEAE. Erect or suberect herbs with modified *haplostele*. *Prothalli* generally subterranean, saprophytic, with symbiotic fungus. *Embryo* often living a subterranean life for some years.

Only British genus :   **Lycopodium.**

† Haplostele. The cauline vascular cylinder in its simpler form, in which the centre is occupied by xylem, and this is typically surrounded by phloem, pericycle and endodermis (Brebner, *Ann. Bot.* 1902, p. 523 and Tansley, *Evolution of Filicinean Vascular System*, 1908).

Suborder 2. LIGULATAE. Heterosporous forms with ligule.

Family SELAGINELLACEAE. Erect or suberect herbs. Haplostelic or polystelic. *Rhizophores* often arising exogenously and bearing endogenous *roots. Leaves* rarely spiral, more often in two dorsal and two ventral rows. *Sporophylls* aggregated in *cones*, which contain both *macro-* and *microsporophylls. Macrosporangia* typically with four *macrospores, microsporangia* with numerous *microspores. Embryo* with two cotyledons.

Only genus : Selaginella.
(The British species is haplostelic with spirally arranged leaves.)

Order 2. **Isoëtales.** Submerged subaquatic or terrestrial plants with very short unbranched fleshy *stems*, bearing alternating whorls of sterile and fertile, elongated, subulate or filiform *leaves. Stele* with anomalous secondary thickening. *Sporangia* bulky, single, in pit on adaxial face of leaf base adjoining the *ligule. Macrosporangia* with numerous *macrospores, microsporangia* with very numerous *microspores.* Sterile *trabeculae* between the spores. *Spermatozoids* multiciliate.

Only family : ISOËTACEAE.
Only genus : Isoëtes.

Class 3. FILICARIAE. *Leaves* almost always large in relation to the *stem*, often highly compound, mostly with circinate *vernation. Stem* haplostelic, solenostelic†

---

† Solenostele. A stele in which the vascular tissue is arranged in a hollow cylinder of xylem lined within and without by phloem, pericycle and endodermis, and the continuity of which is interrupted by leaf-gaps which do not overlap (Tansley, *l.c.*).

or dictyostelic†, departure of leaf-traces in all but haplostelic forms making definite *leaf-gap* in the stele. *Sporangia* on the undersides or on the edges of ordinary foliage leaves, or of modified leaves (*sporophylls*) with reduced laminae; in the latter case the sporophylls are not aggregated on special shoots, or special regions of the shoot. *Spermatozoids* multiciliate.

Order I. **Filicales.** *Sporangia* mostly arising from single cells, more rarely from groups or from several layers of cells, usually in definite aggregates (*sori*). *Prothallus* not subterranean, generally flat.

Suborder I. FILICINEAE. Wall of the sporangium typically with a prominent cell row or cell group (*annulus*) concerned with the dehiscence of the sporangium. *Spores* all alike producing comparatively large free-living monoclinous or diclinous prothalli.

SIMPLICES (Bower). *Sporangia* grouped in sori each containing few sporangia of simultaneous origin on a flat or nearly flat receptacle or spread over the lower surface of the fertile leaf, relatively large, sessile or subsessile, with a large spore output. *Annulus* transverse or oblique or (as in the British genus) represented by a group of thickened cells. *Indusium* typically absent.

Family OSMUNDACEAE. *Stem* upright covered with the bases of spirally arranged leaves. *Stele* with continuous mantle of peripheral *phloem* and anastomosing *xylem strands* surrounding a *pith*. *Petiolar vascular*

† Dictyostele. A stele in which the hollow cylinder of the solenostelic type is interrupted by leaf-gaps which overlap (Brebner, *l.c.*).

strand horseshoe-shaped in cross section. *Sporangia* massive, shortly stalked with vertical dehiscence, not arranged in *sori*, but thickly set on the surface of special, often reduced, fronds or pinnae.
Only British genus: **Osmunda.**

GRADATAE (Bower). *Sporangia* basipetal, on a more or less elongated receptacle, of moderate size, shortly stalked, with medium spore output. *Annulus* oblique. *Indusium* a basal cup, entire or 2-lipped.

Family HYMENOPHYLLACEAE. *Rhizomes* creeping, haplostelic, bearing fronds with thin filmy laminae which have no stomata or intercellular spaces in the mesophyll. *Sori* marginal on the ends of excurrent veins.
Prothalli filiform. Indusium tubular. **Trichomanes.**
Prothalli flat. Indusium 2-lipped.
                                   Hymenophyllum.

MIXTAE (Bower). *Sporangia* of various ages mixed without regular orientation, grouped in *sori*, on flat or nearly flat receptacles, more rarely scattered over the leaf, typically small, with thin elongated stalks and with small spore output. *Annulus* vertical, incomplete, dehiscence transverse. *Indusium* various or absent.

Family POLYPODIACEAE. *Rhizome* of various habit, typically dictyostelic.

Tribe 1. *Woodsieae.* *Sori* globose, often sub-marginal. *Indusium* cup-shaped, opening laterally or splitting irregularly. *Spores* bilateral.
Indusium often ciliate, attached basally. **Woodsia.**
Indusium hooded, attached laterally. **Cystopteris.**

Tribe 2. *Aspidieae.* *Sori* circular or subcircular in outline, flattish ; *indusium* attached centrally, peltate.

Sorus cordate, indusium reniform, attached in the sinus or absent. **Dryopteris** (including **Phegopteris**).

Sorus circular, indusium attached in the centre.

Polystichum.

Tribe 3. *Asplenieae.* *Sori* elongated to linear, developed along veins from which the lateral *indusium* springs.

Subtribe 1. *Blechninae.* *Sori* on veins parallel to the midrib of pinna or pinnule of special fertile leaves.

Only British genus : **Blechnum.**

Subtribe 2. *Aspleninae.* *Sori* on lateral veins of pinna or pinnules of ordinary fronds.

Indusium with single flap entire. **Asplenium.**

Indusium with single flap, margin lacerate.

Athyrium.

Indusium with two flaps, one on either side of sorus.

Scolopendrium.

Tribe 4. *Pterideae.* *Sori* elongated to linear, developed along special veins. *Indusium* mostly absent, sometimes developed as an infolding of the leaf margin.

Subtribe 1. *Gymnogramminae.* *Sori* occupying the whole of the fertile veins.

Only British genus : **Gymnogramme.**

Subtribe 2. *Cheilanthinae.* *Sori* occupying the distal ends of the fertile veins.

Only British genus : **Cryptogramme.**

Subtribe 3. *Adiantinae.* *Sori* occupying the distal ends of the fertile veins, within the revolute leaf margin.

Only genus : **Adiantum.**

Subtribe 4. *Pteridinae. Sori* on intramarginal vein-anastomoses, covered by the indusial leaf margin.
Only British genus (sorus with an additional inner indusial ciliate flap): **Pteridium** (= **Pteris** auct.).

Tribe 5. *Polypodieae. Leaves* separating from rhizome by definite absciss-layers. *Indusium* absent.
Only British genus (sori circular): **Polypodium.**

Suborder 2. HYDROPTERIDINEAE. Wall of sporangium without *annulus. Spores* of two kinds (heterosporous). *Macrospores* borne singly in macrosporangia, and each developing a female prothallus. *Microspores* numerous in each microsporangium, and developing minute few-celled male prothalli.

Family 1. MARSILIACEAE. *Rhizome of sporophyte* solenostelic with two rows of leaves. *Sori* enclosed in special organs (*sporocarps*) which arise as branches of the leaves. *Female prothallus* producing a single *archegonium.*
Only British genus (leaves filiform): **Pilularia.**

Family 2. SALVINIACEAE. *Rhizome of sporophyte* floating, haplostelic, with two or three rows of partly submerged leaves, on which the *sori*, enclosed in *indusia*, are borne. *Female prothallus* producing several archegonia.
Leaves in two rows, deeply bilobed. True roots present†. ***Azolla.**

Order 2. **Ophioglossales.** *Fertile leaves* with one or more simple or pinnate branches bearing the bulky,

† Cf. the exotic genus *Salvinia* which is rootless.

marginal sporangia which arise from groups of cells. *Prothallus* tuberous, subterranean with numerous sunken antheridia and archegonia.

Family OPHIOGLOSSACEAE. *Stem* short, vertical, sunk in the earth and sending up a few leaves which take several years to develop. Sterile leaf-segment pinnate. Pinnules with digitate nerves. Fertile segments branched. **Botrychium.** Sterile leaf-segment simple, entire with reticulate veins. Fertile segment unbranched, spike-like.

Ophioglossum.

Division 2. **EMBRYOPHYTA SIPHONOGAMA** (*Phanerogamae auct.*).

Alternation of generations obscured by seed formation. *Sporophyte* cormophytic, very various, heterosporous.

The *megaspore* (*embryo sac*) is not discharged from the *sporangium* (*ovule*) which after fertilisation ripens to form a *seed.*

Microspore (*pollen grain*) on germination forming a *tube. Antheridium* represented by a single cell which divides to form two naked *generative cells* which in the lowest Gymnosperms become transformed into *Spermatozoids.*

Subdivision I. **GYMNOSPERMAE.** *Microsporangia* on under surface of sporophylls. *Carpels* not fused together to form an ovary, thus no *stigma* is present.

The *microspores* (*pollen grains*) germinate at the *micropile. Seeds* naked on the ovuliferous scales or on a direct prolongation of the axis.

Prothallus (*endosperm*) formed before fertilisation.

Class 1. **CONIFERAE.** *Stem* much branched. No *vessels* in secondary wood. *Leaves* mostly narrow, linear or lanceolate. *Flowers* diclinous. *Perianth* o or bracteoid. *Cotyledons* 2–15, free.

Fam. 1. TAXACEAE. *Carpels* solitary or few. *Seed* drupaceous exserted.

Subfamily *Taxoideae. Stamens* with 3–8 *pollen sacs. Carpels* with two ovules or the ♀ *flower* reduced to a single *ovule.*

Tribe *Taxeae.* ♀ *flower* reduced to a solitary terminal *ovule.*
Only British genus (Mesophyll with no resin canals):
Taxus.

Fam. 2. PINACEAE. *Resin canals* always present in leaves. *Micro-* and *megasporophylls* numerous, arranged in cone-like *flowers. Seeds* not exserted.

Tribe 1. *Abietineae. Leaves* spiral. *Carpels* divided into *cover-scales* and *ovuliferous scales*, the latter with two ovules side by side on the upper surface.
1. Only long shoots present with spiral evergreen leaves.
Bark greyish white. Leaves flat, emarginate with two white lines beneath. Cones erect, scales deciduous.
*Abies.
Bark reddish. Leaves 4-angled in section, acute. Cones pendulous, scales persistent. ***Picea.**
As *Picea* but leaves flat. ***Tsuga** (including ***Pseudo-tsuga** with long 3-fid cover-scales).

2. Long and short shoots present.
Long and short shoots both with deciduous leaves.
Cone-scales broad, coriaceous, persistent. ***Larix.**
Long and short shoots both with evergreen leaves.
Cone-scales deciduous. ***Cedrus.**
Long shoots with scales, short ones with 2–5 evergreen
leaves. Cone-scales woody, persistent. **Pinus.**

Tribe 2. *Cupressineae. Leaves* opposite or whorled.
Ovules erect.
Cone-scales fleshy and confluent in fruit. Only
British genus : **Juniperus.**

Subdivision 2. **ANGIOSPERMAE.** *Ovules* enclosed in an *ovary* formed of coherent *carpels* or of
one carpel with coherent margins. *Stigma* present to
receive pollen.

Class 1. **MONOCOTYLEDONES.** *Embryo* with
one cotyledon. *Mature stem* with closed vascular
bundles. *Parallel venation* predominant. *Flowers* often
3-merous.

A. **Orders with preponderating inconstancy
in the number of floral leaves.** (For B see page 23.)

*a. Typical achlamydeous (not apopetalous) flowers
still occur.* (For *b* see page 21.)

α. *Achlamydeous flowers predominate. Great inconstancy in
number of stamens and also of carpels.* (For β see page 11.)

Order 1. **Pandanales.** Marsh herbs (or tropical
trees and lianes). *Leaves* linear. *Inflorescence* cylindrical
or spherical. *Flowers* naked or with haplochlamydeous,
bracteoid perianth, diclinous, ♂ with ∞ –1 *stamens*,
♀ with ∞ –1 *carpels. Seed* with endosperm.

Fam. 1. TYPHACEAE. Herbs with rhizomes and distichous, linear leaves. *Inflorescence* cylindrical, ♀ *flowers* below, ♂ above. ♂ *flowers* of 2–5 stamens often united ; *pollen* in tetrads. ♀ *flowers* of one *carpel* on an elongated axis which bears hairs. *Stigma* narrow. *Ovule* one, pendulous. *Fruit* an achene. *Seed* with thin perisperm and fleshy endosperm.

Only genus :   Typha.

Fam. 2. SPARGANIACEAE. Herbs. *Leaves* distichous. *Inflorescence* globose. *Lower inflorescences* ♀, *upper* ♂. *Flowers* homochlamydeous ; *tepals* 3–6. *Stamens* 3–6. *Carpels* (1–2) each with one pendulous ovule. *Stigmas* 1–2, lanceolate. *Fruit* a drupe. *Seed* with mealy endosperm.

Only genus :   Sparganium.

β. *Naked flowers still present: but in this order there occur all possible gradations from Achlamydy to Heterochlamydy, also from Hypogyny to Epigyny. Sporophylls definite or indefinite.* (For a see page 10, for γ see page 15.)

Order 2. **Alismatales** (Fluviales, Helobiae). Herbaceous water or marsh plants with *scales* (*Squamulae intravaginales*) in the leaf axils. *Flowers* cyclic or hemicyclic ; achlamydeous, haplochlamydeous or diplochlamydeous (homochlamydeous or heterochlamydeous); hypogynous, or epigynous. *Stamens* ∞ – 1. *Carpels* ∞ –1, usually apocarpous when superior. *Endosperm* absent or scanty.

Suborder 1. POTAMOGETONINEAE. *Flowers* hypogynous, achlamydeous or haplo- to homochlamydeous.

Fam. 1. POTAMOGETONACEAE. Submerged or floating fresh or salt water plants. *Leaves* mostly distichous. *Flowers* mostly small, solitary or in spikes, diclinous or monoclinous with 4–1-merous whorls.

12 Potamogetonaceae *Potamogetonaceae, Naiadaceae*

Perianth mostly absent. *Stamens* 4–1. *Carpels* 4–1, each with one pendulous ovule. *Fruit* often drupaceous.

a. Flowers in spikes.

Tribe 1. *Zostereae. Flowers* diclinous in a flat, dorsiventral spike enclosed in the sheath of the uppermost foliage leaf. *Pollen* filiform.

Only British genus: **Zostera.**

Tribe 2. *Potamogetoneae. Axis of spike* terete, *flowers* monoclinous.

Spike usually many-flowered. Tepals 4 : stamens 4 : achenes sessile. **Potamogeton.**

Spike 1–2-flowered. Tepals 0 : stamens 2 : achenes stipitate. **Ruppia.**

b. Flowers solitary or cymose, diclinous.

Tribe 3. *Zannichellieae. Perianth* absent or very simple. *Carpels* 4–3.

Only British genus : **Zannichellia.**

Fam. 2. NAIADACEAE. Slender, submerged fresh or brackish water herbs. *Stem* with central cylinder of elongated cells enclosing a *canal. Leaves* opposite, linear, toothed, almost imbricating. *Flowers* diclinous. ♂ of two cup-like envelopes and a terminal *anther.* ♀ of one cup-like envelope (which may be absent) and one *carpel* containing one basal anatropous *ovule.*

Only genus : **Naias.**

*Fam. 3. APONOGETONACEAE. Herbs with tuberous sympodial stems. *Leaves* submerged or lamina floating. *Spike* aerial enclosed in deciduous *spathe. Flowers* monoclinous. *Perianth* 3–1 petaloid. *Stamens* 6 or more. *Carpels* 3–6. *Styles* 3–6. *Fruit* leathery with 2 or ∞ seeds.

Only genus : ***Aponogeton.**

Fam. 4. SCHEUCHZERIACEAE (Juncaginaceae).
Scapigerous marsh herbs with narrow, rush-like *leaves*.
Inflorescence racemose. *Flowers* monoclinous or diclinous,
4–1-merous. *Perianth* homochlamydeous, herbaceous.
Each carpel with 1–2 anatropous ovules.

Tribe. *Triglochineae. All flowers* of inflorescence
similar.
Bracts absent. Anthers subsessile, short. **Triglochin.**
Inflorescence bracteate. Filaments and anthers long.
Scheuchzeria.

Suborder 2. ALISMATINEAE. *Flowers* hypogynous,
mostly heterochlamydeous. *Ovules* on the ventral
suture.

Fam. 5. ALISMATACEAE. Marsh herbs. *Leaves*
chiefly radical. *Inflorescence* often much branched.
Flowers mostly monoclinous and heterochlamydeous,
3-merous. *Stamens* 6–∞. *Carpels* 6–∞ each with
1–∞ anatropous ovules. *Styles* 6–∞. *Fruit* of achenes
or follicles.
  a. Flower-axis flat. Stamens 6 in one whorl.
Carpels also more or less cyclic.
  α. Carpels with one ovule.
Leaves erect, flowers whorled. **Alisma.**
Leaves floating, flowers subsolitary. **Elisma.**
  β. Carpels with two or more ovules.
Leaves erect, carpels connate. **Damasonium.**
  b. Flower-axis convex. Stamens mostly more
than 6, sometimes spiral. Carpels arranged in a head,
acyclic.
  Flowers hermaphrodite. Fruit ribbed, scarcely compressed. **Echinodorus.**

Flowers diclinous. Fruit strongly compressed.
Sagittaria.

Suborder 3. BUTOMINEAE. *Flowers* hypogynous
or epigynous, mostly heterochlamydeous. *Ovules* scattered over inner surface of carpel on branching placentae.

Fam. 6. BUTOMACEAE. Marsh plants with conspicuous *flowers. Inflorescence* cymose, often umbel-like.
Flowers monoecious, usually markedly heterochlamydeous, 3-merous. K 3, C 3, A 6 (3 pairs) + 3–∞, in the
last case the outer ones sterile. G 6–∞ often united
at the base with ∞ ovules scattered over their inner
surfaces. *Fruit* of follicles.
Only British genus (perianth homochlamydeous,
persistent):   **Butomus.**

Fam. 7. HYDROCHARITACEAE. Submerged or
floating water plants. *Leaves* mostly spiral, sometimes
cyclic or distichous. *Flowers* solitary or cymose, at
first enclosed in an envelope of 1–2 bracts, usually
diclinous, mostly heterochlamydeous, 3-merous. K 3,
C 3, A 3 + (3 + 3 + 3 + 3), the inner and outer stamens
sometimes staminodal. G (2–15). *Placentae* parietal,
bearing ∞ orthotropous to anatropous ovules with
2 integuments. *Stigma* often deeply bifid. *Fruit*
mostly dehiscing irregularly with ∞ seeds.

  a. Carpels 6–15. *Placentae* projecting far inwards.

Subfamily 1. *Stratiotoideae.* Fresh water plants.
Foliage leaves spiral.

  Tribe 1. *Stratioteae. Leaves* partly submerged.
♀ *flowers* sessile in the 2-leaved spathe.
  Leaves erect narrow serrate.   **Stratiotes.**

Graminaceae 15

Tribe 2. *Hydrochariteae. Leaves* floating. ♀ *flowers* stalked in the spathe.

Leaves orbicular. **Hydrocharis.**

b. Carpels usually 3. *Placentae* not projecting far inwards.

Subfamily 2. Vallisnerioideae. Fresh water plants. *Perianth* often small. *Stigma* short.

Tribe. Hydrilleae. Leaves 1-nerved, in whorls. ♂ *flowers* 1-3 in a spathe.

Leaves linear, 3 in a whorl. *Elodea.

γ. *Naked flowers dominant. Stamens rarely, and carpels never, indefinite.* (For α see page 10, for β see page 11.)

Order 3. **Glumiflorae.** Grassy herbs with sheathing *leaves. Flowers* naked, more rarely with perianth of bristles†, in the axils of imbricating bracts (*glumes*). *Gynaecium* always 1-locular with one ovule.

Fam. 1. GRAMINACEAE. Herbs (grasses). *Stems* usually terete. *Leaves* distichous, sheath usually split to the base and furnished with *ligule* or hairs at junction of sheath and lamina. *Flowers* in the axils of glumes with 2-nerved '*palea*' opposite glume, usually monoecious, naked. The lowermost glumes are usually without flowers (*empty*). Deeply bifid scale (*lodiculae*) often present opposite palea. *Stamens* mostly 3. *Anthers* versatile. *Stigmas* usually 2, feathery. *Fruit* a caryopsis‡. *Seed* with copious endosperm. *Embryo* with peltate enlargement (*scutellum*) of cotyledon in whose distal

† *Oreobolus*, a South American genus of *Cyperaceae*, tribe *Rhynchosporeae*, has a perianth of glume-like tepals.
‡ A caryopsis is a superior achene in which the pericarp and testa are united.

hollow lie the plumule and the radicle, the latter surrounded by a sheath (the *Coleorhiza*).

Subfamily 1. *Panicoideae*. *Spikelets* 1-flowered with axis (*rachilla*) not produced beyond flower, rarely 2-flowered, in which case the lower flower is imperfect, when ripe falling off as a whole. (See footnote.)

a. *Hilum* point-like, *spikelets* dorsally compressed or terete.

Tribe 1. *Paniceae*. *Flowering glumes* and *paleae* harder than *empty glumes*. *Lower empty glume* usually smaller.

Pedicels of spikelets naked or hairy. ☉**Panicum.**
Pedicels of spikelets with stiff bristles. ☉**Setaria.**

b. *Hilum* linear, *spikelets* laterally compressed.

Tribe 2. *Oryzeae*. *Empty glumes* minute or 0. *Stamens* often 6 (but 3 in Brit. Sp.).

Empty glumes absent. Only British genus : **Leersia.**

Subfamily 2. *Poëoideae*. *Spikelets* 1–many-flowered, when 1-flowered the *axis* (*rachilla*) is often prolonged beyond the flower, mostly jointed distally to the *empty glumes* which remain behind when the *flowering glumes* fall away†. When 2–many-flowered there are always distinct internodes between the flowers.

a. *Spikelets* with distinct stalks, arranged in panicles, spike-like panicles or racemes. (For β see page 19.)

Tribe 3. *Phalarideae*. *Spikelets* 1-flowered, with 4 *empty glumes*. *Palea* 1-nerved.

3rd and 4th glumes empty, awnless, reduced to small scales. **Phalaris.**

---

† In Alopecurus, Polypogon, Holcus and Spartina the spikelets fall as a whole.

3rd and 4th glumes empty, small with dorsal awns.
Anthoxanthum.
3rd and 4th glumes, or at any rate the 3rd with a triandrous ♂ flower, almost as long as the 1st and 2nd. Upper flowers hermaphrodite, diandrous. **Hierochloë.**
Tribe 4. *Agrostideae. Spikelets* 1-flowered with two *empty glumes. Palea* 2-nerved.
* Rachilla not produced beyond flowering glume.
Panicle effuse, fruit included in hardened glume.
Milium.
Panicle dense, cylindric, spikelets falling as a whole, flowering glume with dorsal, bent awn. **Alopecurus.**
Panicle dense, cylindric, empty glumes persistent, flowering glume awnless. **Phleum.**
Spikelets dorsally compressed in a simple spike.
Mibora.
Panicle loose, flowering glume small, membranous.
Agrostis.
Panicle contracted, spikelets falling as a whole, empty glumes with awns. **Polypogon.**
Rachilla with long silky hairs, empty glumes acuminate. **Calamagrostis.**
** Rachilla produced beyond the flowering glume.
Empty glumes large, dilated at base, much larger than the minute 4-toothed flowering glume. **Gastridium.**
Empty glumes large, flowering glumes bifid, with slender awns. **Apera.**
Empty glumes large, flowering glumes with awns, rachilla ciliate. **Deyeuxia.**
Spikelets large in dense panicle, rachilla long and silky. **Ammophila.**
Empty glumes with feathery hairs, flowering glume 3-awned. **Lagurus.**

Tribe 5.   *Aveneae.*   *Spikelets* 2–many-flowered.
Flowering glumes usually shorter than the two *empty glumes*, usually with dorsal, rarely apical, bent and twisted awn.

A.   Spikelets falling as a whole.   **Holcus.**

B.   Flowering glumes in fruit becoming detached from the persistent empty glumes.

α.   Rachilla not produced.
Flowering glumes awned, 2-toothed.   **Aira.**

β.   Rachilla produced.
Flowers 2, awn bent in middle, tip clavate.

Corynephorus.

Flowers 2, awn straight, acute.   **Deschampsia.**

Flowers 2–6, flowering glumes deeply bifid with twisted awns.   **Trisetum.**

Flowers 2–6, flowering glumes entire or 2-toothed with long awns.   **Avena.**

Flowers 2, upper hermaphrodite, lower ♂.

Arrhenatherum.

Tribe 6. *Festuceae.* As *Aveneae* but *flowering glumes* usually longer than the two *empty glumes.* *Awn* absent, or apical and straight.

α. Spikelets 2 or more flowered : rachilla not bearded. Flowering glumes with three broad teeth.

Sieglingia.

β. Spikelets 2 or more flowered : rachilla with silky hairs.   **Phragmites.**

γ. Spikelets subspicate or capitate, with imperfect spikelets on pedicels below them.
Imperfect spikelets ciliate.   **Sesleria.**
Imperfect spikelets with stiff bristles.   **Cynosurus.**

δ. Spikelets 2 or more flowered. Flowering glumes
1 or 3-nerved, all alike.

Spikelets arranged in a spike-like panicle. Flowering
glumes scarious. **Koeleria.**

Spikelets conical, terete, in a slender panicle.
Flowering glumes cartilaginous. **Molinia.**

Spikelets in effuse panicle, branches whorled. Flowering glumes coriaceous. **Catabrosa.**

ε. Spikelets 2 or more flowered. Flowering glumes
3–5-nerved, upper empty, convolute. **Melica.**

ζ. Spikelets 3–many-flowered. Flowering glumes
5 or more nerved.

Spikelets few-flowered, clustered in secund panicles.
Dactylis.

Spikelets panicled, drooping. Glumes broad, obtuse.
Briza.

Spikelets panicled. Flowering glumes compressed,
keeled, tips nerved, without awns. **Poa.**

Spikelets very many flowered. Flowering glumes
convex, obtuse, tips without nerves or awn. **Glyceria.**

Spikelets in racemes or spike-like panicles. Flowering glumes convex, tip nerved, acute or with awn. Ovary
glabrous. **Festuca.**

Spikelets many-flowered, in panicles. Flowering
glumes convex. Ovary tip villous. **Bromus.**

Spikelets subsessile, distichous. Flowering glumes
convex. Ovary tip villous. **Brachypodium.**

β. *Spikelets* in two rows which approach one another
forming a unilateral spike or raceme with unjointed
axis. (For α see page 16.)

Tribe 7. *Chlorideae. Spikelets* 1–many-flowered.
Empty glumes and *paleae* mostly 2-nerved.

Spikelets falling as a whole. Spikes solitary.
              Spartina.

Empty glumes persistent. Spikes digitate. **Cynodon.**

γ. *Spikelets* in two opposite rows. (For α see page 16, for β see page 19.)

  Tribe 8. *Hordeae.* *Spikelets* sessile in notches on a simple *rachis* forming a *spike.*

  α. Style 1. Spikes unilateral.   **Nardus.**

  β. Styles 2. Spikes bilateral.

  I. Spikelets solitary in notches of rachis.

  † Spikelets with the narrow sides to rachis. **Lolium.**

  †† Spikelets inserted broadside to the rachis.

  Spikelets 1–2-flowered in a slender spike. Spikelets sunken in the notches.   **Lepturus.**

  Spike stout. Spikelets not sunken in the notches.
      Triticum (incl. **Agropyrum** and ***Secale**).

  II. 2–6 spikelets in each notch.

   1. Spikelets 1-flowered.   **Hordeum.**

   2. Spikelets 2–many-flowered. **Elymus.**

  Fam. 2. CYPERACEAE. Grass-like herbs often with triangular stems. *Leaves* with closed sheaths. *Ligule* usually absent. *Flowers* in spikelets or spikelet-like cymes, monoclinous or diclinous, naked or rarely with homochlamydeous perianth. *Stamens* mostly 3–1, rarely more. *Anthers* basifixed. *Carpels* (3–2). *Styles* 3–2 with filiform *stigmas.* *Fruit* an achene with free *seed.*

  Subfamily 1. *Scirpoideae.* *Spikelets* with many hermaphrodite flowers. Individual flowers in the spikelet may be diclinous. *Perianth* present or absent.

  Tribe 1. *Scirpeae.* No *bracteoles.*

Subtribe 1. *Cyperinae. Glumes* distichous.
Only British genus: **Cyperus.**

Subtribe 2. *Scirpinae. Glumes* spiral.
Spikelets solitary, terminal. Bristles 3–8, included.
Style thickened at base.   **Heleocharis.**
Spikelets clustered and lateral. Bristles 0, or 3–8,
included. Style not thickened at base.   **Scirpus.**
Spikelets solitary or clustered, terminal. Bristles
very long, cottony.   **Eriophorum.**

Subfamily 2. *Rhynchosporoideae. Flowers* in
spikelet-like cymes (*pseudo-spikelets*) which are arranged
in spikes or heads, mono- or diclinous. *Perianth bristles*
present or absent.

Tribe 2. *Rhynchosporeae. Pseudo-spikelets* few-
flowered.
Pseudo-spikelets terete. Bristles slender or absent.
Fruit beaked.   **Rhynchospora.**
Pseudo-spikelets compressed; glumes distichous.
Fruit not beaked.   **Schoenus.**
Pseudo-spikelets terete. Bristles 0. Fruit obtuse.
                            Cladium.

Subfamily 3. *Caricoideae. Flowers* always naked,
diclinous, in many-flowered (rarely few-flowered), mono-
or diclinous spikelets. ♀ *flower* enclosed in modified
bracteole (*utriculus*).
Spikelets 1–2-flowered.   **Kobresia.**
Spikelets many-flowered.   **Carex.**

*b. Achlamydeous flowers rare: their occurrence is
mostly due to reduction and is correlated with spathe-
development. Definite number of stamens and carpels*

dominant, but numerous stamens and more than three carpels often occur. (For *a* see page 10.)

Order 4. **Spathiflorae.** Mostly sympodial, rarely forming erect stems. (The *Lemnaceae* are free floating and undifferentiated.) *Flowers* cyclic, haplochlamydeous or diplochlamydeous, homochlamydeous or naked, 3–2-merous, hermaphrodite or diclinous, often very reduced, always inserted on a *spadix* more or less enveloped by a bract (*spathe*). *Floral bracts* absent.

Fam. 1. ARACEAE. Mostly herbs. *Rhizome* often tuberous. *Leaves* often with netted veins. *Flowers* mono- or diclinous, rarely dioecious, 2–3-merous, or reduced sometimes to a single *stamen* or *carpel*. *Fruit* usually a berry. *Seed* with fleshy outer integument.

Subfamily 1. *Pothoideae.* Land plants. No *latex* or *raphides* present. *Leaves* distichous or spiral. Lateral nerves of 2nd and 3rd order usually netted. *Flowers* mostly hermaphrodite.

Tribe 1. *Acoreae.* *Leaves* not differentiated into stalk and lamina. *Perianth* present. *Ovules* orthotropous.
Only British genus : ⊙Acorus.

Subfamily 2. *Aroideae.* Mostly tuberous land or marsh plants with netted leaves. *Latex* present. *Flowers* diclinous. *Perianth* usually absent.

Tribe 2. *Areae.* *Spadix* mostly with vestigial flowers. *Gynaecium* of one carpel.
Only British genus : **Arum.**

Fam. 2. LEMNACEAE. Small free-swimming thalloid plants not differentiated into *stem* and *leaves*. *Flowers* monoecious. ♂ flower of one *stamen*. ♀ of one *carpel* with 1–6 basal *ovules*.

Subfamily 1. *Lemnoideae*. *Roots* present. *Inflorescence* with spathe and two ♂ flowers.

Roots several. Spirodela.
Root solitary. Lemna.

Subfamily 2. *Wolffoideae*. *Roots* absent. *Inflorescence* without spathe and with only one ♂ flower.
Wolffia.

B. **Orders with typically pentacyclic flowers. Whorls typically isomerous, mostly 3-merous.** (For A see page 10.)

a. *Flowers homochlamydeous to heterochlamydeous, very rarely naked. Bracteoid perianths still occur but petaloid perianths predominate. Hypogyny and Actinomorphy preponderate.* (For b see page 29.)

Order 5. **Farinosae.** Mostly herbs, rarely with well-developed stem. *Flowers* cyclic, homo- or heterochlamydeous, 3 or 2-merous. General floral formula $T 3 + T 3$ (more rarely $K 3 + C 3$), $A 3 + 3$, $G (3)$. One whorl, or all but one of the stamens may be absent. *Ovules* orthotropous, but anatropous ovules also occur. *Endosperm* mealy.

Fam. ERIOCAULACEAE. Perennial, scapigerous herbs. *Leaves* narrow, chiefly radical. *Flowers* in involucrate heads, monoclinous, usually monoecious. *Perianth* membranous. Outer whorl of stamens mostly absent. *Carpels* (2–3). *Styles* 2–3. *Ovary* 2–3-loc, each loculus with a pendulous orthotropous ovule. *Fruit* a loculicidal capsule.

Subfamily *Eriocauloideae. Stamens* 4 or 6. *Perianth-segments* free, inner ones with gland near apex.

Only British genus : **Eriocaulon.**

Order 6. **Liliiflorae.** Characters of *Farinosae*, but *ovules* mostly anatropous, and *seeds* usually with fleshy or cartilaginous endosperm. Exceptionally 2 or 4-merous flowers occur.

Suborder 1. JUNCINEAE. *Perianth* homochlamydeous, not petaloid. *Endosperm* containing starch.

Fam. 1. JUNCACEAE. Herbs, usually perennating by underground sympodial rhizomes. *Inflorescence* compound, usually many-flowered. *Flowers* homochlamydeous, 3-merous, mostly monoclinous. *Perianth* bract-like. The inner whorl of stamens sometimes not developed. *Carpels* (3). *Style* 1 with three filiform *stigmas. Ovary* 1 or 3-loc with 1 or ∞ ovules. *Capsule* loculicidal. *Embryo* straight.

Annual or perennial. Glabrous. Capsules 3-loc, many-seeded. **Juncus.**

Perennial. Hairy. Capsule 1-loc, 3-seeded. **Luzula.**

Suborder 2. LILIINEAE. *Perianth* mostly petaloid, homo- very rarely heterochlamydeous. Inner circle of *stamens* present. *Endosperm* with no starch.

Fam. 2. LILIACEAE. *Flowers* usually homochlamydeous and monoclinous. *Perianth* usually petaloid, *segments* free or united. *Stamens* 6 (3 in Ruscus). *Ovary* usually superior, mostly 3-loc with axile *placentae. Fruit* various.

Subfamily 1. *Melanthioideae. Rhizome* or *corm.*
Inflorescence terminal. *Anthers* extrorse and *capsule*
septicidal, or *anthers* introrse and *capsule* septicidal, or
anthers extrorse and *capsule* loculicidal, rarely *anthers*
introrse and *capsule* loculicidal. *Fruit* never a berry.

 a. Rhizome. Seeds long, flat and winged, or angled.

Tribe 1. *Tofieldieae. Leaves* sessile, distichous.
Stamens 6. *Anthers* introrse. *Style* divided or 0.
Flowers yellow. Style very short, stigma small.
Capsule loculicidal. **Narthecium.**
Flowers greenish-white. Style absent, stigmas 3,
short. Capsule septicidal. **Tofieldia.**

 b. Corm, or short *rhizome. Seeds* subglobose.

Tribe 2. *Colchiceae. Leaves* radical. *Scape* short
subterranean with 1–3 *leaves. Anthers* introrse. *Capsule*
septicidal.
Perianth tube long. Styles separate. **Colchicum.**

Subfamily 2. *Asphodeloideae. Rhizome. Leaves*
usually radical. *Inflorescence* mostly terminal. *Anthers*
introrse, sometimes opening at apex. *Capsule* loculicidal.
Only British genus: **Simethis.**

Subfamily 3. *Allioideae. Bulb* or short *rhizome.*
Inflorescence a cymose umbel enclosed by two broad,
sometimes united, leaves, more rarely subtended by two
narrow bracts, or reduced to a single flower.

Tribe 3. *Allieae. Bulb,* or stem with thickened
base. *Perianth* free or united. *Stamens* 6 or only
3 fertile.
Inflorescence few-flowered, subtended by two free
bracts. **Gagea.**

Inflorescence usually many-flowered, subtended by two coriaceous, mostly united, bracts.   **Allium.**

Subfamily 4. *Lilioideae. Bulb. Inflorescence* terminal, racemose. *Anthers* introrse. *Capsule* loculicidal.

Tribe 4. *Tulipeae. Bulb,* imbricated or tunicated. *Stem* bearing several foliage leaves, rarely only one. *Flowers* few, axillary, or solitary and terminal.

α. Anthers versatile.

Flowers large. Nectary median or obscure. *Lilium.

β. Anthers basifixed.

Flowers large, drooping, segments not reflexed. Nectary oblong.   **Fritillaria.**

As *Fritillaria* but flowers erect and nectary absent.
Tulipa.

Flowers few, small. Perianth-segments spreading.
Lloydia.

Tribe 5. *Scilleae. Tunicated bulb. Aerial stem* without leaves. *Flowers* in the axils of bracts.

1. Perianth-segments free or nearly so.

Filament filiform or flattened at base.   **Scilla.**

Filaments flattened throughout.   **Ornithogalum.**

2. Perianth-segments united.

[Corolla funnel-shaped, not contracted, all flowers fertile.   *****Hyacinthus.]**

Corolla globose, upper flowers imperfect.   **Muscari.**

Subfamily 5. *Asparagoideae. Rhizome* subterranean, ending in flowering shoots, or monopodial with lateral flowering shoots. *Fruit* a berry.

a. *Flowers* homochlamydeous.

Tribe 6. *Asparageae.* *Rhizome* ending in leafy stems. *Cauline leaves* small, scale-like, subtending narrow or broad leaf-like shoots (*Phylloclades*).

Stem herbaceous. Flowers axillary. Stamens 6. Filaments distinct. **Asparagus.**

Stem shrubby. Flowers on the phylloclades dioecious. Stamens 3 with connate filaments. **Ruscus.**

Tribe 7. *Polygonateae.* As *Asparageae,* but *stem* with large, broad *foliage leaves.*

Leaves many. Flowers axillary. Perianth tubular 6-cleft. **Polygonatum.**

Leaves 2. Flowers in terminal racemes. Perianth 4-partite. **Maianthemum.**

Tribe 8. *Convallarieae.* *Rhizome* monopodial with lateral *inflorescence-axes.*

Subtribe *Convallariinae.* Style ending in one small stigma. **Convallaria.**

b. Flowers heterochlamydeous.

Tribe 9. *Parideae. Foliage leaves* net-veined forming a whorl. *Flowers* solitary or umbellate. *Fruit* a berry.

Leaves 4 or more in a whorl. Flowers 4 or more merous. Only British genus: **Paris.**

Fam. 3. AMARYLLIDACEAE. Resembling, in essential points, Liliaceae. *Anthers* mostly introrse. A *corona,* formed of staminal stipules, is often present. *Ovary* mostly inferior. *Placentation* axile. *Ovules* anatropous in two rows on each placenta. *Fruit* a loculicidal capsule or berry. *Seeds* usually few.

Subfamily 1. *Amaryllidoideae. Bulb. Scape* leafless with a solitary flower, or involucrate umbellate inflorescence.

Tribe 1. *Amaryllideae.* No *corona.*

Subtribe *Galanthinae.* Perianth actinomorphic, without tube.   Loculi of ovary with ∞ ovules.
Perianth-segments all equal.   **Leucojum.**
Outer perianth-segments larger than inner.
                                                ☉**Galanthus.**

Tribe 2. *Narcisseae. Corona* present as tube, ring or scales.

Subtribe *Narcissinae.* Stamens included in the cuplike corona.   **Narcissus.**

Fam. 4. DIOSCOREACEAE. Lianes. *Rhizome* often tuberous. *Leaves* often sagittate, veins reticulate. *Flowers* in racemes, mostly diclinous, homochlamydeous. *Perianth* sepaloid, usually with short tube.   G (3), 3 or 1-loc with axile or parietal *placentae,* each placenta mostly with two superposed, anatropous ovules. *Styles* 3, sometimes bifid.   *Fruit* a capsule or berry.

Tribe *Dioscoreae. Flowers* diclinous. *Ovules* 2 in each loculus.
Only British genus (berry imperfectly 3-celled):
                                                Tamus.

Suborder 3. IRIDINEAE. As Liliineae, but *inner whorl of stamens* aborted.

Fam. 5. IRIDACEAE. Mostly herbs with equitant leaves and terminal inflorescences. *Flowers* hermaphrodite, homo- or heterochlamydeous 3-merous, actinomorphic or zygomorphic. *Stamens* always 3 (the outer whorl). *Anthers* extrorse.   G ($\overline{3}$) with many anatropous ovules on axile placentae†. *Styles* 3, often divided and leaf-like. *Capsule* loculicidal with globose or angular seeds.

Subfamily 1. *Crocoideae.* Small herbs often with underground flower stalks. *Flowers* solitary, or several axillary around a terminal flower.

Scape short. Perianth tube short. **Romulea.**
Scape 0. Perianth tube long. ⊙**Crocus.**

Subfamily 2. *Iridoideae. Stem* distinct. *Leaves* equitant. *Flowers* several in each spathe.

Tribe 1. *Moraeeae. Perianth tube* short or absent. *Stigmas* on underside of leaf-like style-arms. *Capsule* not enclosed in spathe.
Only British genus: **Iris.**

Tribe 2. *Sisyrinchieae.* As above, but *style-arms* mostly terete.
Only British genus: ⊙**Sisyrinchium.**

Subfamily 3. *Ixioideae.* As *Iridoideae*, but only *one flower* enclosed in each *spathe.*

Tribe *Gladioleae. Flowers* strongly zygomorphic, often curved.
Only British genus: **Gladiolus.**

b. Flowers homochlamydeous to heterochlamydeous but in the first case the perianth is petaloid. Epigyny throughout. Zygomorphy dominant. (For *a* see page 23.)

Order 7. **Microspermae.** *Flowers* cyclic, homo- or heterochlamydeous, 3-merous, typically diplostemonous but *androecium* often shows marked reduction.

† The Mediterranean genus *Hermodactylus* has a 1-loc ovary with parietal placentae. In all other respects this monotypic genus is very near to *Iris.*

Ovary inferior, 3 or 1-loc with numerous, minute ovules. *Endosperm* present or absent.

Suborder GYNANDRAE. *Flowers* always zygomorphic, no *endosperm*.

Family ORCHIDACEAE. Perennial herbs. *Flowers* homo- or heterochlamydeous, typically 3-merous, almost invariably monoclinous, zygomorphic, usually resupinate by twisting of the ovary so that the posterior petal occupies an anterior position and is often developed into a labellum. Of the *stamens* usually only the anterior† one of the outer whorl is fertile. In subfamily Diandrae the two lateral stamens of the inner whorl are usually fertile. Frequently (e.g. in Orchis) the two anterior stamens of the inner whorl are represented by staminodes on the sides of the column. *Pollen* cohering in masses (*Pollinia*). *Carpels* (3̄) sunk in the floral axis which is prolonged beyond the insertion of the perianth into a *column* which bears the stamens. *Stigmas* 3, usually on the surface of the column, the posterior one rudimentary or developed into a beak-like structure (*rostellum*) beneath the anther or between its cells. *Ovary* mostly 1-loc with 3 parietal placentae and ∞ ovules. *Fruit* a capsule with very numerous, minute seeds without *endosperm*. *Embryo* not, or only slightly, differentiated.

Subfamily 1. *Pleonandrae* (Diandrae). The two lateral (rarely all 3) stamens of the inner whorl fertile. All three stigmas similar and receptive.

† The terms *anterior* and *posterior* in this description refer to position before the occurrence of resupination.

Tribe 1. *Cypripedileae.* *Flowers* zygomorphic. *Column* curved towards the slipper-shaped labellum. Only British genus: **Cypripedium.**

Subfamily 2. *Monandrae.* *Anterior stamen* of outer whorl fertile. *Lateral stigmas* receptive, odd one rudimentary or developed as rostellum.

a. Basitonae. *Anther* confluent with column. *Pollinia* with appendages at base which are in connection with glands in rostellum.

Tribe 2. *Ophrydeae.* The only tribe. Terrestrial orchids with root tubers.

Subtribe *Serapiadinae.* *Column* short with labellum at base. *Anthers* erect. *Stigma* disc-like. *Glands of pollinia* enclosed in rostellar pouch.

Spur long: both glands in a single pouch. **Orchis.**

Spur absent: both glands in one pouch. **Aceras.**

Spur absent: glands in two distinct pouches. **Ophrys.**

Subtribe *Gymnadeniinae.* As above, but *pollen masses* naked or surrounded by processes of the anthers.

Without spur. **Herminium.**

With spur. **Habenaria.**

b. Acrotonae. *Anther* deciduous. *Pollen masses* without appendages.

(*α*) Acranthae: *Inflorescences* terminal on branches whose basal portions make up a sympodial shoot. [(*β*) Pleuranthae (*main axis* sympodial but *flowers* borne on special lateral axes): not represented in Britain.]

I. Convolutae. *Leaves* convolute in bud. *Sheath* and *blade* not distinct from one another. *Pollen* granular. *Anther* withering without falling.

Tribe 3. *Neottieae* (only tribe).

Subtribe 1. *Cephalantherinae. Labellum* with distinct proximal lobe (*hypochilium*) which is often spurred. *Anthers* erect. *Rostellum* short or absent.

α. Stem leafy: no spur.

Flowers in racemes, ovary straight.   **Epipactis.**

Flowers in spikes, ovary twisted.   **Cephalanthera.**

β. Stem scaly, without foliage leaves, labellum spurred.

Labellum turned upwards, column short. **Epipogon.**

Subtribe 2. *Spiranthinae. Leaves* soft, net-veined. *Anther* as long as the beaked rostellum and lying against it. *Pollinia* not divided into distinct masses.

A. Median sepal and petals forming a helmet surrounding the base of the labellum.

Spike spirally twisted.   **Spiranthes.**

B. Sepals and petals standing apart.

Scape with two opposite leaves.   **Listera.**

Scape brown, leafless.   **Neottia.**

Subtribe 3. *Physurinae.* As above but *pollinia* divided into numerous distinct masses.

Rhizome creeping, leaves several.   **Goodyera.**

II. Duplicatae. *Leaves* folded in bud.

Tribe 4. *Liparideae. Pollinia* 4, waxy, without *appendages.*

A. Foliage leaves present.

Column short, straight. Anther persistent. **Malaxis.**

Column slender, bent forwards. Anther deciduous.   **Liparis.**

B. No foliage leaves. Roots absent. Rhizome branched, coral-like.   **Corallorhiza.**

Class 2. **DICOTYLEDONES.** *Embryo* usually with two cotyledons. *Stem* with open bundles. *Leaves* usually with reticulate venation. *Flowers* often 4–5-merous.

Subclass 1. **ARCHICHLAMYDEAE.** *Perianth* simple or absent. *Flowers* either achlamydeous, or haplochlamydeous, or diplochlamydeous with polypetalous corolla. Sympetaly rare (marked in *Cotyledon*). Apopetaly occurs not infrequently.

a. **Amentiflorae** (Moss)†. **Mostly trees or shrubs.** ♂ **flowers usually, and** ♀ **flowers often, in catkins. Perianth absent or haplochlamydeous and bracteoid.** (For *b* see p. 37.)

Order 1. **Salicales.** Deciduous trees or shrubs. *Leaves* alternate, stipulate, seldom deeply lobed. *Flowers* dioecious, both sexes in simple catkins. *Perianth* cup- or saucer-shaped, or modified into one or two or rarely more nectaries. *Stamens* 2–∞, filaments free or rarely connate. *Carpels* (2). *Ovary* 1-locular with ∞ ovules on two parietal placentae. *Ovules* anatropous with two integuments. *Fruit* a loculicidal capsule quite free from the bracts. *Seeds* small with basal tuft of hairs. *Endosperm* 0.

Only family : SALICACEAE.

† Engler has no names for these groups, simply referring to them as *a, b, c* and *d*. The names used in this book are those made use of by Dr Moss in his lectures at Cambridge.

C.   3

Leaves usually broad. Bracts laciniate. Perianth cup-like or saucer-like, more or less oblique. **Populus.** Leaves usually narrow. Bracts entire. Perianth of usually separate nectaries. **Salix.**

Order 2. **Myricales.** Shrubs or trees. Leaves alternate, simple, exstipulate, sweet scented. *Flowers* achlamydeous, diclinous, monoecious or dioecious, sometimes with bracteoles at the base. *Stamens* usually 4. *Carpels* (2). *Ovary* 1-locular with one basal orthotropous ovule with one integument. *Styles* 2, filiform. *Fruit* a drupe with waxy surface often adnate to bracteoles. *Seeds* without endosperm.

Fam. MYRICACEAE.
Only British genus :   **Myrica.**

Order 3. *****Juglandales.** Trees with alternate, exstipulate, pinnate leaves. *Flowers* diclinous, monoecious, achlamydeous or haplochlamydeous. *Stamens* 3–40. *Carpels* (2). *Ovary* inferior, 1-locular with one basal, orthotropous ovule with one integument. *Fruit* a drupe or nut with adherent bracts. *Endosperm* 0.

Fam. *****JUGLANDACEAE.** ♀ *flower* with perianth which is fused with the ovary and also with the bract and bracteoles.

*****Juglans.**

Order 4. **Fagales.** Trees or shrubs. *Leaves* alternate, simple, with stipular, usually caducous bud-scales. *Flowers* homochlamydeous or achlamydeous, mostly diclinous and monoecious. ♂ *flowers* in catkins. ♀ *flowers* in catkins or small spikes, or 1–3 seated in an involucre of free or connate bracts. *Stamens* often opposite the tepals. *Carpels* (2–6). *Ovary* sub-inferior, usually more or less completely 2–3-locular (after fertilisation) with 1–2 pendulous, anatropous, parietal ovules in each loculus. *Fruit* usually a nut. *Endosperm* 0.

Fam. 1. BETULACEAE. Flowers usually appearing before leaves. *Perianth* present in flowers of only one sex. ♂ *catkins* compound. ♂ *flowers* united to their bracts. *Stamens* 2–10. *Filaments* often dividing into two, each branch bearing a half-anther. ♀ *catkins* often minute. *Carpels* (2). *Styles* 2.

Tribe 1. *Coryleae.* ♂ *flowers* without perianth. ♀ *flowers* with perianth. *Fruit* not winged, enclosed in the enlarged, herbaceous bracts.

Leaves in bud folded parallel to lateral nerves. ♀ spike large. Fruiting bracts open 3-lobed. **Carpinus.**

Leaves in bud folded parallel to midrib. ♀ spike minute. Fruiting bracts forming a cupule. **Corylus.**

Tribe 2. *Betuleae.* ♂ *flowers* in dichasia, with perianth. ♀ *flowers* without perianth. *Fruit* usually winged.

Filaments branched. ♀ catkins falling at end of the first summer. **Betula.**

Filaments simple. Empty ♀ catkins persistent for more than one year. **Alnus.**

Fam. 2. FAGACEAE. *Flowers* and *leaves* appearing together, or flowers appearing after leaves. *Flowers* of both sexes with perianth of 4–7 connate tepals. ♂ *catkins* simple or compound. *Stamens* 4–14. ♀ *flowers* in 3–2-flowered dichasia or solitary. *Carpels* usually 3–4. *Styles* 3–4. *Fruit* surrounded by involucre (*cupule*).

3—2

Tribe 1. *Fageae.* ♂ *catkins* globose. Two angular
nuts in each cupule.
Only British genus :   **Fagus.**
Tribe 2. *Castaneae.* ♂ *catkins* elongate. *Nuts*
rounded.
♂ catkins erect. Nuts 3, enclosed in spinous cupule.
                                      ☉**Castanea.**
♂ catkins pendulous. Nut single, exserted. **Quercus.**

Order 5. **Urticales.** Herbs, shrubs or trees
with stipulate, often roughly hairy, leaves. *Inflorescence*
mostly cymose, but catkins still occur (*Urtica*). *Flowers*
cyclic, homochlamydeous, rarely naked. *Stamens* opposite the tepals. *Carpels* 2–1, superior. *Ovary* usually
1-locular, with one ovule with two integuments.

Fam. 1. ULMACEAE. Woody plants without latex.
Leaves often distichous, simple, oblique at base, strongly
pinnately nerved, serrate, with rough hairs and deciduous
stipules. *Flowers* mostly hermaphrodite. *Perianth*
4–5-lobed. *Stamens* as many. *Filaments* straight in
bud. *Carpels* (2). *Styles* 2. *Ovule* usually solitary,
pendulous and anatropous. *Fruit* a nut or drupe.
Endosperm usually o.

Subfamily *Ulmoideae. Flowers* in clusters. *Pedun-
cles* in the axils of scales. *Fruit* never drupaceous.
Embryo straight.
Only British genus (fruit broadly winged): **Ulmus.**

Fam. 2. CANNABACEAE. Aromatic herbs without
latex. *Leaves* palmately nerved, and lobed or divided.
Stipules persistent. *Flowers* dioecious. ♂ *flowers, tepals 5,
stamens 5.* ♀ *flowers* with perianth of low entire mar-

gined cup surrounding the *gynaecium*. *Carpels* 2. *Styles* 2. *Ovule* pendulous. *Fruit* a nut. *Endosperm* fleshy. *Embryo* curved.

Stem twining. Leaves cordate, 3–7-lobed. **Humulus.**
Stem erect. Leaves palmate.   ***Cannabis.**

Fam. 3. URTICACEAE. Mostly herbs without latex. *Stinging hairs* often present. *Flowers* diclinous. *Perianth* of both sexes usually of four free tepals. *Stamens* as many incurved in bud. *Carpel* 1. *Fruit* a nut enclosed by inner two or by all the tepals. *Endosperm* 0. *Embryo* straight.

Tribe 1. *Urereae*. *Stinging hairs* present. *Perianth* of ♀ flowers 4-partite.

Only British genus (leaves opposite, stigma brush-like):   **Urtica.**

Tribe 2. *Parietarieae*. No *stinging hairs*. ♀ *perianth* (or involucre ?) tubular.

Only British genus :   **Parietaria.**

b. **Petaloideae** (Moss). **Flowers haplochlamydeous, perianth often petaloid. Diplochlamydeous flowers rare (Rumex).** (For *a* see page 33, for *c* see page 40.)

Order 6. **Santalales.** Partial or entire parasites. *Leaves* entire, exstipulate. *Flowers* cyclic, homochlamydeous with *stamens* opposite *tepals*. *Carpels* $(\overline{2-3})$, rarely $(\overline{1})$. One *ovule* to each carpel, pendulous from apex of loculus or from free central placenta, or placenta and ovules not differentiated. One or no integuments.

Suborder 1. SANTALINEAE. *Ovules* differentiated from placenta, often without integument.

Fam. SANTALACEAE. Chlorophyllous hemi-parasites. *Leaves* mostly alternate, entire, exstipulate. *Tepals* 4–5, valvate in bud, bracteoid or petaloid, united below. *Stamens* adnate to tepals. *Ovary* 1-locular with a columnar, free central placenta from which are suspended the 1–3 ovules which are without integument. *Fruit* indehiscent, with one endospermous seed.

Tribe *Thesieae*. *Perianth* epigynous. *Tube* long. Only British genus :   **Thesium.**

Suborder 2. LORANTHINEAE. *Ovules* mostly not differentiated.

Fam. LORANTHACEAE. Evergreen shrubs, mostly parasitic on trees. *Leaves* thick, usually opposite and exstipulate. *Tepals* usually 4, valvate, fleshy. *Floral axis* united to *gynaecium*, often forming a crenate ring (*calyculus*). *Ovary* 1-loc, usually with only one fertile embryo sac. *Fruit* a pseudo-berry, inner layer of axis becoming sticky.

Subfamily *Viscoideae*. *Flowers* diclinous. No distinct *calyculus*.

Tribe *Visceae*. *Flowers* single or in groups in the axils of persistent bracts. *Placenta* basal. Only British genus :   **Viscum.**

Order 7. **Aristolochiales.** *Flowers* cyclic, homochlamydeous, epigynous, actinomorphic or zygomorphic. *Perianth* petaloid. *Ovary* inferior, usually 3–6-locular. *Ovules* always ∞, mostly axile.

Fam. ARISTOLOCHIACEAE. Herbs or lianes, with alternate, long-stalked, usually cordate or reniform, exstipulate leaves. *Flowers* usually hermaphrodite, actinomorphic or zygomorphic. *Perianth* petaloid, with three lobes or one lip. *Stamens* usually 6 or 12, free or united to the style. *Carpels* ($\overline{4-6}$) with numerous axile anatropous ovules with two integuments. *Fruit* a capsule. *Seed* with small embryo in copious endosperm.

Tribe 1. *Asareae.* Herbs with reniform leaves. *Flowers* actinomorphic.
Only genus (perianth campanulate): ⊙**Asarum.**

Tribe 2. *Aristolochieae.* *Flowers* zygomorphic.
Only British genus (perianth tubular, dilated at base): ***Aristolochia.**

Order 8. **Polygonales.** *Leaves* alternate, usually with sheathing bases. *Flowers* haplochlamydeous to heterochlamydeous, actinomorphic. *Ovary* superior, 1-locular, with one basal orthotropous ovule with two integuments.

Fam. POLYGONACEAE. Herbs, rarely shrubs or trees. *Leaves* usually entire with scarious or fleshy amplexicaul, stipular sheath. *Flowers* small in compound inflorescences, cyclic or partially spiral, usually hermaphrodite. *Tepals* 3–6 free. *Stamens* 5–9. G($\underline{3}$) or ($\underline{2}$) with as many *styles*. *Fruit* a nut. *Seed* with copious mealy endosperm. *Embryo* straight or curved.

Subfamily 1. *Rumicoideae.* *Flowers* cyclic. *Endosperm* not fissured.

Tribe *Rumiceae.* *Leaves* with conspicuous sheath. Stamens in one whorl. Fruit not winged. **Rumex.** Stamens in two whorls. Fruit winged.

Rheum (including **Oxyria**).

Subfamily 2. *Polygonoideae.* *Flowers* mostly spiral. *Endosperm* not fissured.

Tribe *Polygoneae.* Herbs, more rarely undershrubs. Only British genus (tepals free, all erect in fruit):
Polygonum (including ***Fagopyrum**).

c. (**Consisting of the one order Centrospermae.**) **Corolla usually haplochlamydeous, sepaloid or petaloid. Heterochlamydy, however, is not uncommon.** (For *d*, Heterochlamydeae, see page 45, for *b* page 37.)

Order 9. **Centrospermae.** Mostly herbs, often with peculiar secondary thickening. *Leaves* often entire, usually exstipulate. *Stipules* scarious if present. *Flowers* homo- or heterochlamydeous. *Stamens* as many as and opposite to perianth segments or $\infty - 1$. *Carpels* $\infty - 1$, usually united. *Ovary* mostly superior and 1-loc with $1-\infty$ campylotropous ovules. *Placentation* basal or free central. Two integuments present. *Embryo* curved. *Seeds* often reniform and with granulate testa. Perisperm often present.

Suborder 1. CHENOPODINEAE. *Perianth* homochlamydeous and sepaloid. *Tepals* not more than 5 with *stamens* opposite them. *Ovule* solitary.

Fam. 1. CHENOPODIACEAE. Herbs, rarely shrubs, often with 'bladder-hairs' ('mealiness'). *Leaves* alternate, exstipulate, fleshy, entire or irregularly lobed, sometimes much reduced. *Flowers* small, homochlamydeous. *Tepals* 5, 3, or 1, valvate. *Stamens* as many or fewer opposite the tepals, bent inwards in bud. *Carpels* (2), rarely (3–5). *Ovary* 1-loc with one basal ovule. *Fruit* a nutlet or pyxidium enclosed in persistent perianth.

A. *Cyclolobeae. Embryo* annular or horse-shoe shaped.

Tribe 1. *Beteae. Flowers* hermaphrodite. *Stamens* 1–5, united below. *Ovary* subinferior. *Stigma* short, broad, papillose within. *Fruit* opening with lid.
Only British genus: **Beta.**

Tribe 2. *Chenopodieae. Flowers* mostly hermaphrodite. *Stigma* linear. *Fruit* indehiscent.
Only British genus: **Chenopodium.**

Tribe 3. *Atripliceae. Flowers* usually diclinous. ♂ with *perianth* and without *bracteoles.* ♀ without *perianth* and with persistent *bracteoles.*
Only British genus: **Atriplex.**

Tribe 4. *Salicornieae. Flowers* hermaphrodite embedded in the leaves.
Only British genus: **Salicornia.**

B. *Spirolobeae. Embryo* spirally twisted.

Tribe 5. *Suaedeae. Leaves* succulent, glabrous. *Bracteoles* small. *Style-arms* papillose all round.
Only British genus: **Suaeda.**

Tribe 6. *Salsoleae. Leaves* often with filiform hairs. *Bracteoles* as large or larger than tepals. *Style-arms* papillose within.
Only British genus: **Salsola.**

*Fam. 2. AMARANTACEAE. Very near *Chenopodiaceae* but *perianth* not herbaceous, but scarious and coloured ; the *tepals* often end in awns.
*Amarantus.

Suborder 2. *PHYTOLACCINEAE. *Flowers* haplo-or heterochlamydeous with a tendency to spiral arrangement of parts. *Stamens* sometimes numerous. *Carpels* sometimes only slightly united.

Fam. 3. *AIZOACEAE. Usually herbs, often with fleshy leaves. *Stipules* absent or scarious. *Flowers* usually haplochlamydeous. *Perianth segments* free or united, sepaloid. *Stamens* 5 or more, the outer ones sometimes represented by petaloid staminodes. *Carpels* (2–∞). *Ovary* 2–∞-loc, rarely 1-loc. *Placentation* basal or free central (parietal in *Mesembryanthemum*).

Tribe *Mesembryanthemeae. Ovary* inferior. *Staminodes* present or not.
Staminodes very numerous, petaloid.
                                        *Mesembryanthemum.

Suborder 3. PORTULACINEAE. *Flowers* heterochlamydeous. *Sepals*† 2. *Petals* 4–5.

Fam. 4. PORTULACACEAE. Herbs with entire fleshy leaves and scarious stipules. *Flowers* hermaphrodite, heterochlamydeous. *Sepals* 2. *Petals* 4–5, free or connate below, deciduous. *Stamens* 5–∞. *Carpels* (3–5) with 2–∞ ovules on a basal placenta. *Fruit* capsular.
A. Stamens 3–5. Ovary superior.
Petals free, stamens 5.   *Claytonia.
Petals connate, stamens 3.   Montia.
B. Stamens numerous. Ovary subinferior.
                                        *Portulaca.

† These structures may also be regarded as bracteoles. In the N. Am. genus *Lewisia* they are 4–8 in number, subulate, and placed on the scape below the insertion of the perianth.

Suborder 4. CARYOPHYLLINEAE. *Flowers* heterochlamydeous with equal number of sepals and petals, cyclic throughout, sometimes apopetalous.

Fam. 5. CARYOPHYLLACEAE. Herbs, rarely shrubs, with opposite, rarely spiral, entire leaves. *Stipules* o or scarious. *Flowers* mostly in dichasia, cyclic, usually heterochlamydeous, 5 rarely 4-merous. *Stamens* usually in two whorls. *Carpels* (5–2) with 1–∞ ovules on a free central or basal placenta. *Fruit* usually capsular.

Subfamily 1. *Alsinoideae. Sepals* free. *Stamens* often perigynous. *Styles* free or united.

a. Petals usually conspicuous. Capsule opening with teeth.

Tribe 1. *Alsineae. Stipules* o. *Styles* free.
Cymes lax. Petals bifid. Capsules globose. **Stellaria.**
Cymes usually lax. Petals bifid. Capsules cylindric.
Cerastium (including Moenchia).
Cymes corymbose. Petals jagged. Capsules cylindric.
Holosteum.
Petals entire. Styles 3–4 (carpels fewer than sepals).
Arenaria.
Petals entire or o. Styles 4–5 (carpels isomerous).
Sagina.

Tribe 2. *Sperguleae. Stipules* scarious. *Styles* free.
Petals white. Carpels and styles 5. **Spergula.**
Petals white or pink. Carpels and styles 3.
Spergularia.

Tribe 3. *Polycarpeae. Stipules* scarious. *Styles* united below.
Only British genus : **Polycarpon.**

b. Petals inconspicuous or absent. Fruit indehiscent.

44 *Caryophyllaceae*

Tribe 4. *Paronychieae. Stipules* scarious. *Petals* minute, bristle-like. *Ovules* 1–4.

A. Sepals obtuse, flat, green.

Leaves linear, opposite and spiral. Stigmas 3.

Corrigiola.

Leaves ovate-oblong, opposite. Stigmas 2.

Herniaria.

B. Sepals awned, laterally compressed, white.

Illecebrum.

Tribe 5. *Sclerantheae. Stipules* 0. *Petals* usually 0. *Ovules* 1–2.

Only British genus (leaves opposite, connate at base): **Scleranthus.**

Subfamily 2. *Silenoideae. Sepals* united. *Petals* and *stamens* hypogynous. *Gynophore* often present. *Styles* free.

Tribe 6. *Lychnideae. Calyx* with commissural ribs.

A. Fruit a capsule.

Carpels isomerous, alternating with sepals. Petals without appendages. Claw 2-winged. **Agrostemma.**

Carpels oligomerous, or if isomerous opposite sepals. Capsule with septa. Teeth as many as styles.

Viscaria.

As *Viscaria*, but teeth twice as many as styles.

Silene.

Capsule with no septa. Teeth as many as styles.

Lychnis.

As *Lychnis*, but teeth twice as many as styles.

Melandryum.

B. Fruit a berry. ***Cucubalus.**

Tribe 7. *Diantheae. Calyx* with no commissural ribs. Bracts immediately beneath capsule. Petals without appendages. Embryo straight.

Dianthus (including **Tunica†**). Bracts not immediately beneath capsule. Petals with appendages. Embryo curved. ***Saponaria.**

d. **Heterochlamydeae** (Moss). **Orders with predominant heterochlamydy.** (For *c* see page 40.)

a. *Apocarpy and Hypogyny predominant.* (For β see page 48.)

Order 10. **Ranales.** *Flowers* spiral, spirocyclic, or cyclic, haplo- to heterochlamydeous, hypogynous to epigynous, actinomorphic or zygomorphic. *Stamens* usually ∞. *Carpels* ∞ –1, usually free.

Suborder 1. NYMPHAEINEAE. *Flowers* predominantly spiral. *Ovules* usually scattered over the inner surface of the carpels but sometimes solitary at the apex (*Ceratophyllaceae*).

Fam. 1. NYMPHAEACEAE. Aquatic herbs. *Leaves* floating, often peltate. *Flowers* solitary spiral, spirocyclic or cyclic, homo- or heterochlamydeous, hermaphrodite, actinomorphic. *Flower-axis* convex, or hollow and united to gynaecium. P 6–∞, A 6–∞, G 3–∞ free or united, each with 1–∞ ovules on their inner surface. *Ovules* with two integuments. *Seeds* often with aril. *Cotyledons* thick.

Subfamily *Nymphaeoideae. Flowers* spirocyclic. *Carpels* ∞, with ∞ ovules completely covering the septa. *Endosperm* and *perisperm* present.

† An ill-defined genus differing from *Dianthus* in having smaller flowers and broad, membranous sepaline commissures.

Tribe 1. *Nuphareae.* K 6–12, C ∞. Fruit a berry.
Flowers globose, yellow.   **Nuphar.**

Tribe 2. *Tetrasepaleae.* K 4, C ∞.
Flowers expanded, white, blue, or red. **Nymphaea.**

Fam. 2. CERATOPHYLLACEAE. Submerged aquatics
with whorls of transparent, forked leaves, which later
become cartilaginous. *Flowers* solitary in the leaf axils,
haplochlamydeous, diclinous, monoecious. ♂ *perianth*
about 12. *Stamens* 12–16. ♀ *perianth* about 9–10.
Carpel 1 with one apical, orthotropous ovule with one
integument. *Style* long, subulate.
Only genus :   **Ceratophyllum.**

Suborder 2. RANUNCULINEAE. *Flowers* spiral to
cyclic. *Ovules* on the ventral suture.

Fam. 3. RANUNCULACEAE. Mostly acrid herbs.
Leaves often divided. *Flowers* rarely completely cyclic,
haplo- or heterochlamydeous, in first case with petaloid
perianth between which and androecium honey-leaves
are often present, hermaphrodite, actinomorphic or zygomorphic. *Stamens* usually numerous, free. *Carpels* ∞ –1
usually free with ∞ –1 ovules. *Fruit* of one-seeded
achenes or many-seeded follicles, or a berry (*Actaea*).
Endosperm copious.

A. *Ovules* on both sides of ventral suture, rarely
solitary. *Fruit* a follicle with ∞ –1 seeds.

*Tribe 1. *Paeonieae.* *Flowers* usually solitary, without
honey-leaves. *Carpels* fleshy. *Outer integument* longer
than inner.
Only genus :   *Paeonia.

Tribe 2. *Helleboreae. Flowers* 1 or several. *Carpel wall* rarely fleshy. *Outer integument* not longer than inner.

I. Honey-leaves without spur. Sometimes absent.
A. Leaves undivided or merely lobed. Honey-leaves absent. **Caltha.**
B. Leaves deeply lobed or compound. Honey-leaves present.
† Leaves palmately lobed or compound.
α. Honey-leaves flat with naked honey-pits.
Trollius.
β. Honey-leaves tubular, at least at base.
Leaves palmately divided, sepals membranous, deciduous. ***Eranthis.**
Leaves pedate. Sepals herbaceous, persistent.
Helleborus.
†† Leaves doubly ternate or pinnate, fruit a berry.
Actaea.

II. Honey-leaves with spur.
1. Flowers actinomorphic with five honey-leaves.
Aquilegia.
2. Flowers zygomorphic with two honey-leaves.
Honey-leaves sessile, dorsal sepal spurred.
*Delphinium.
Honey-leaves stalked, dorsal sepal hooded.
☉**Aconitum.**

B. *Ovule* solitary at base of ventral suture. *Achene* one-seeded.

Tribe 3. *Anemoneae* (only tribe).
Shrubs. Leaves opposite. Sepals valvate. **Clematis.**
Herbs. Leaves spiral. Sepals 4–5, inconspicuous. Honey-leaves o. **Thalictrum.**

Herbs.   Leaves spiral.   Scape with involucre of 3 leaves.   Sepals 4–20.   Honey-leaves 0.   **Anemone.**
Herbs.   Leaves spiral.   Sepals 5–8.   Honey-leaves 5–16, conspicuous.   ***Adonis.**
Herbs.   Leaves spiral.   Sepals 5, spurred.   Honey-leaves small, tubular.   Flower-axis ultimately elongate.
                                                         Myosurus.
Herbs.   Leaves spiral.   Sepals 3–5, not spurred. Honey-leaves 5, flat, petaloid.   **Ranunculus.**

Fam. 4.   BERBERIDACEAE.   Herbs, shrubs, or trees. *Leaves* simple or compound.   *Flowers* solitary or racemose, cyclic, homo- or heterochlamydeous, hermaphrodite 3–2-merous, actinomorphic.   *Perianth* with 2–4 whorls, two whorls of honey-leaves often present.   *Stamens* in two whorls, anthers opening by lids or valves. *Carpel* 1, rarely several, with ∞–1 ovules on ventral suture or basal.   Two integuments.   *Fruit* a berry.   *Seed* with endosperm.

Subfamily *Berberidoideae.*   *Leaves* pinnate, or terminal leaflet alone present.   *Honey-leaves* present.

Tribe 1.   *Berberideae.   Inflorescence* on lateral short shoots.
Only British genus (flowers 3-merous, fruit a berry):
                                                         Berberis.

*Tribe 2.   Epimedieae.   Inflorescence terminal.
Only British genus (flowers 2-merous.   Petals with discolorous appendages, fruit dehiscent):
                                                  *Epimedium.

β.   *Syncarpy and Hypogyny dominant.*   (For α see page 45, for γ see page 54.)

Order 11. **Rhoeadales.** Usually herbs. *Flowers*
often racemose, cyclic (except androecium in some
cases), usually heterochlamydeous, hypogynous, actinomorphic or zygomorphic. *Carpels* (∞ –2). *Ovules* with
two integuments.

Suborder 1. RHOEADINEAE. *Flowers* usually heterochlamydeous. *Sepals* usually 2.

Fam. 1. PAPAVERACEAE. Herbs, often with white
or coloured latex. *Flowers* hermaphrodite, usually
nodding in bud. *Sepals* 2. *Petals* usually 4. *Stamens* ∞,
or if only 4–2, then branched from the base. *Carpels*
(2–16) with parietal *placentae* and ∞ *ovules*, or with one
basal ovule. *Fruit* usually capsular. *Embryo* minute.
Endosperm oily.

Subfamily 1. *Papaveroideae. Petals* without spur.
Stamens ∞. *Carpels* 2–∞.

Tribe 1. *Chelidonieae. Latex* yellowish or reddish.
Style ending in two undivided branches which alternate
with the *placentae. Stigmatic surface* within and on the
edges of these branches.
Only British genus: ⊙**Chelidonium.**

Tribe 2. *Papavereae. Latex* yellow or white.
Stigmas lying above the *placentae.*
a. Fruit long, dehiscing to the base.
Fruit with septum.   **Glaucium.**
Fruit with no septum. Flowers violet. *****Roemeria.**
b. Fruit elongate, ovoid or globose, dehiscent above.
Stigmas 4–5, style distinct.   **Meconopsis.**
Stigma subsessile, on a disc.   **Papaver.**

C.   4

Subfamily 2. *Fumarioideae.* *Flowers* mostly transversely zygomorphic. One of two *outer petals* gibbous or spurred. *Stamens* opposite them, each stamen divided from the base into three. Ovules numerous, fruit a 2-valved capsule.

Corydalis.

Ovules 2, fruit 1-seeded, indehiscent. **Fumaria.**

Suborder 2. CAPPARIDINEAE. *Flowers* heterochlamydeous. *Sepals* 4 or more.

Fam. 2. CRUCIFERAE. Annual or perennial herbs usually with unicellular simple or branched hairs. *Leaves* spiral, exstipulate. *Flowers* in bractless racemes, hermaphrodite, actinomorphic. K 2 + 2. C 4. A 2 (short) + 4 (long). G (2). *Placentation* parietal, but *ovary* is 2-loc on account of growth of septum (*replum*) from placentae. *Stigmas* 2 on short style above placentae. *Fruit* mostly capsular. *Endosperm* absent.

Position of cotyledons and radicle as follows:

o || notorhizal (radicle incumbent).

o = pleurorhizal (radicle accumbent).

o > > orthoplocous (cotyledons conduplicate).

The following arrangement of the British Cruciferae is according to de Candolle and is adapted from Hooker's *Students' Flora*, ed. 3. See also *Syllabus*, ed. 7, p. 198, where another arrangement is given.

A. Siliquosae. *Pods* usually much longer than broad, dehiscent throughout their whole length, not compressed at right angles to septum.

Tribe 1. *Arabideae* (Siliquosae pleurorhizae). *Seeds* 1-seriate (2-seriate in Arabis and Nasturtium). *Coty-*

ledons flat. *Radicle* accumbent o $=$. (*Flowers* white, yellow or lilac.)

α. Stigmatic lobes erect, or decurrent to the style.
Matthiola.

β. Stigma small, simple, terminal.
Hairs forked. Lateral sepals pouched. Pods compressed or 4-angled, valves 1-nerved. ***Cheiranthus.**
Hairs simple or o. Pods terete, valves turgid, not
elastic. Seeds minute, 2-seriate. **Nasturtium.**
Pods 4-angled, valves keeled. Seeds oblong. Flowers
yellow. **Barbarea.**
Pods flat, valves 1-nerved, not elastic. Flowers
mostly white. **Arabis.**
Rhizome not fleshy. Pods flat, valves elastic. Funicle
filiform. **Cardamine.**
Rhizome fleshy. Pods flat, valves elastic. Funicle
dilated. **Dentaria.**

Tribe 2. *Sisymbrieae* (Siliquosae notorhizae). *Seeds*
usually 1-seriate. *Cotyledons* flat. *Radicle* incumbent
o ||. (*Flowers* as above.)
Glabrous, or hairs simple and spreading. Flowers
yellow. Sepals equal. Stigma obtuse. **Sisymbrium.**
Hairs appressed, branched. Flowers yellow. Stigma
obtuse. **Erysimum.**
Hairs spreading. Flowers white or lilac, fragrant.
Lateral sepals gibbous. Stigmas decurrent on style.
*Hesperis.

Tribe 3. *Brassiceae* (Siliquosae orthoplocae). *Seeds*
1-2-seriate. *Cotyledons* longitudinally folded o > >.
Radicle incumbent. (*Flowers* yellow.)
Pods terete or angled. Seeds 1-seriate,·globose.
Brassica.

4—2

Pods compressed.   Seeds 2-seriate, compressed.
Diplotaxis.

B.   Siliculosae.   *Pods* short, dehiscent through their whole length.

a. Latiseptae.   Pods compressed parallel to the replum, which is hence as broad as the pod's greatest diameter.

Tribe 4.   *Alysseae* (Latiseptae pleurorhizae).   *Seeds* 2-seriate.   *Radicle* accumbent o =.

Pods oblong, compressed, many-seeded.
Draba (including **Erophila**).

Pods circular, 2–8-seeded.   ***Alyssum.**

Pods globose, many-seeded.   **Cochlearia.**

Tribe 5.   *Camelineae* (Latiseptae notorhizae).   *Seeds* 2-seriate.   *Radicle* incumbent o ||.

Tall herbs, flowers small, cauline leaves auricled.
***Camelina.**

Tribe 6.   *Subularieae* (Latiseptae diplecolobae).   As *Camelineae*, but *cotyledons* curved with lengthened bases turned up in direction of radicle.

Scapigerous water-herb with subulate leaves.
Subularia.

b. Angustiseptae.   *Pod* much compressed at right angles to replum, which is hence narrow.

Tribe 7.   *Thlaspeae* (Angustiseptae pleurorhizae).   *Cotyledons* straight, *radicle* accumbent o =.   (*Flowers* white.)

Pods obovate, notched.   Petals equal.   Filaments without scales.   **Thlaspi.**

Pods orbicular.   Petals white or lilac, very unequal.   Filaments without scales.   **Iberis.**

Pods oblong. Petals white, unequal. Filaments
with scales.   **Teesdalia.**
Pods oblong. Petals equal. Filaments without
scales.   **Hutchinsia.**
Tribe 8. *Lepidieae.* *Cotyledons* straight, incurved or
longitudinally folded. *Radicle* incumbent o | |. (*Flowers*
white.)
Pods many-seeded.   **Capsella.**
Pods 2–4-seeded.   **Lepidium.**
Tribe 9. *Brachycarpeae* (Angustiseptae diplecolobae).
Cotyledons induplicate. *Radicle* incumbent.
Fruit indehiscent, 2-seeded.   **Coronopus.**
C. Nucamentaceae. *Pods* indehiscent, 1-celled,
1-seeded.
Tribe 10. *Isatideae* (Nucamentaceae notorhizae).
Radicle incumbent o | |.
Only British genus: **☉Isatis.**
D. Lomentaceae. *Fruit* with transverse dissepi-
ments, separating into 1-seeded joints.
Tribe 11. *Cakilineae* (Lomentaceae pleurorhizae
o ═). *Radicle* accumbent. *Cotyledons* flat.
Pods compressed, of two dissimilar joints.
Only genus: **Cakile.**
Tribe 12. *Raphaneae* (Lomentaceae orthoplocae
o > >). *Radicle* incumbent. *Cotyledons* longitudinally
folded.
Pods of a lower slender seedless and an upper globose
1-seeded joint.   **Crambe.**
Pods elongate, of 2–several similar joints. **Raphanus.**
Suborder 3. RESEDINEAE. *Flowers* spiro-cyclic,
heterochlamydeous.

Fam. 3. RESEDACEAE. Herbs with spiral leaves. *Flowers* in racemes, spiro-cyclic, hermaphrodite, zygomorphic. *Flower-axis* passing into a short gynophore which expands below the stamens into an eccentric or semilunar disc. K 4–8. C 0–8. A 3–10. G (2–6). *Ovary* 1-locular, open above. *Ovules* usually many on parietal placentae. *Fruit* usually a coriaceous capsule open at the top. No *endosperm*. *Embryo* curved.

Only British genus: **Reseda.**

Order 12. Sarraceniales. Herbs with spiral leaves adapted to insect catching. *Flowers* spiro-cyclic to cyclic, homo- or heterochlamydeous, hypogynous, actinomorphic. *Carpels* (3–5) with parietal or axile placentation and ∞ ovules. *Seeds* minute with endosperm.

Fam. DROSERACEAE. Herbs, mostly without main root. *Leaves* stipulate, spiral, often rosulate; with digestive glands and irritable hairs or tentacles. *Flowers* cyclic, heterochlamydeous, hermaphrodite, actinomorphic. K 5–4. C 5–4. A 5–4–∞. G (5–3). *Styles* 5–3, rarely 1. *Ovary* 1-locular with ∞–3 parietal or basal ovules. *Capsule* mostly 1-celled, loculicidal. *Seeds* ∞–3 with endosperm. *Embryo* minute.

Only British genus: **Drosera.**

γ. *Apocarpy and Hypogyny still occur but Perigyny becomes more frequent. By sinking of the Gynaecium into the hollow flower-axis Syncarpy and Epigynous insertion of the perianth and stamens also takes place.* (For β see page 48, for δ see page 61.)

Order 13. Rosales. *Flowers* cyclic, rarely spirocyclic (*Rosaceae—Rosoideae*), usually heterochlamydeous and actinomorphic. *Carpels* often free. *Placentae*

sometimes thickened, usually with ∞ ovules.—Most of
the families of this order are very difficult to demarcate.

Suborder 1. SAXIFRAGINEAE. *Carpels* as many as
petals or fewer. *Endosperm* mostly copious (scanty in
Crassulaceae).

Fam. 1. CRASSULACEAE. Mostly succulent herbs.
Leaves exstipulate. *Flowers* cyclic, heterochlamydeous,
3–30-merous, haplo- or obdiplostemonous, mostly hermaphrodite and actinomorphic. *Petals* free or united.
Carpels usually isomerous, free or slightly coherent, with
glandular scales at base. *Ovules* mostly 2-seriate on the
ventral suture. *Fruit* of follicles. *Seeds* small oblong
with scanty endosperm.

A. Obdiplostemonous.
* Petals free.
Leaves scattered or 2–3 in a whorl. Flowers 4–5-
merous. **Sedum.**
Leaves rosulate. Flowers 6–20-merous.
*Sempervivum.
** Corolla tubular, 5-lobed. **Cotyledon.**

B. Haplostemonous.
Leaves opposite. Flowers 3–5-merous.
Crassula (including Tillaea).

Fam. 2. SAXIFRAGACEAE. Mostly herbs. *Leaves*
usually spiral. *Stipules* absent or present as outgrowths
from the leaf-sheath. *Flowers* cyclic, usually heterochlamydeous and 5-merous (but *Carpels* mostly oligomerous), hermaphrodite, actinomorphic. *Flower-axis*
convex, flat, or concave, in the latter case usually united
by its whole length to the carpels. *Stamens* usually

obdiplostemonous, but sometimes haplostemonous, more rarely ∞. *Carpels* seldom free and as many as petals, mostly fewer, often two. *Styles* mostly free. *Ovary* 2–1-loc with swollen placentae bearing ∞ ovules in several rows. *Seeds* minute with copious endosperm and small embryo.

Subfamily 1. *Saxifragoideae.* Herbs with spiral leaves. *Flowers* mostly 5-merous. *Carpels* 2, rarely 3–4, free or united below. *Ovules* with two integuments.

Tribe 1. *Saxifrageae. Carpels* rarely free, but *styles* always free.
Flowers 5-merous with corolla. Placentation axile.
Saxifraga.
Flowers 4-merous with no corolla. Placentation parietal. **Chrysosplenium.**

Tribe 2. *Parnassieae. Carpels* (3–4) forming syncarpous gynaecium. *Style* short or absent.
Parnassia.

Subfamily 2. *Ribesoideae.* Shrubs with simple exstipulate *leaves. Flowers* in racemes. *Stamens* 5. *Ovary* inferior, 1-loc, with parietal placentae. *Fruit* a berry.
Only genus: **Ribes.**

Suborder 2. ROSINEAE. *Carpels* ∞ –1. *Ovules* with two integuments. *Endosperm* scanty or absent.

Fam. 3. PLATANACEAE. Trees with bark separating in large scales. *Leaves* spiral, 3–5-lobed, with large connate *stipules. Flowers* in globose heads, diclinous, cyclic, heterochlamydeous, actinomorphic, 3–8-merous, typically isomerous with four alternating whorls but disturbances of this arrangement by abortion occur. *Stamens* with short filaments and clavate anthers with peltate prolongation of connective. *Carpels* free with 1–2 almost orthotropous ovules with two integuments. *Fruit* a caryopsis. *Endosperm* scanty.

*Platanus.

Fam. 4. ROSACEAE. Herbs, shrubs or trees. *Leaves* spiral. *Stipules* sometimes adnate to petiole, rarely absent. *Flowers* cyclic, usually heterochlamydeous, actinomorphic, rarely zygomorphic. *Flower-axis* flat, dish- or cup-shaped: sometimes convex in the middle. *Sepals, petals* and *stamens* on the edge of the axis, perigynous or epigynous. *Petals* often orbicular and concave. *Stamens* mostly 2–4 times as many as sepals, or ∞, rarely 1–5, bent inwards in bud. *Carpels* as many as sepals or 2–3 times as many or ∞, rarely 1–4, free or united with the inner wall of the flower-axis, 1-locular with usually two anatropous ovules. *Styles* apical or on ventral side of carpels. *Fruit* follicular or indehiscent, or drupaceous or a false fruit by union of carpels with flower-axis. *Endosperm* scanty or absent. *Cotyledons* mostly fleshy, plano-convex.

Subfamily *Spiraeoideae. Filaments* narrowed upwards from broad base. *Carpels* 12–1 whorled, rarely sunk into flower-axis or on gynophore. *Ovules* ∞ –2 *Fruit* usually of follicles.

*Tribe *Spiraeeae.* Shrubs rarely herbs. *Seeds* not winged.

*Spiraea†.

† The plants often known in Britain as *S. Ulmaria* and *S. Filipendula* belong more properly to the genus *Ulmaria*, subfamily *Rosoideae*, tribe *Ulmarieae*.

Subfamily *Pomoideae*. *Stipules* distinct. *Carpels* 5–2, united with the inner wall of the hollow flower-axis, usually syncarpous (i.e. united to each other). *False fruit* formed of carpels + axis + lower part of calyx.

A. Carpels free on their ventral sides (i.e. the sides to which the styles are attached).
Fruit a drupe with 3–5 partially exserted stones.
Cotoneaster.

B. Carpels united.
Inner wall of carpel cartilaginous when ripe. **Pirus.**
Inner wall of ripe carpel hard, fruit hence a drupe.
Leaves lobed. Flowers in corymbose cymes.
Crataegus.
Leaves undivided. Flowers solitary. ***Mespilus.**

Subfamily *Rosoideae*. *Carpels* either ∞, on conical gynophore, rarely few not enclosed in axis, or 1–∞ enclosed in persistent flower-axis : each carpel with 1–2 ovules. *Fruit* always indehiscent.

Tribe 1. *Potentilleae*. *Flower-axis* flat. *Stamens* ∞, narrowed upwards from broad base. *Carpels* mostly ∞, on convex gynophore.
No epicalyx. Carpels each with two ovules. Fruit drupaceous. **Rubus.**
Epicalyx present. Carpels with one ovule. Fruit dry, indehiscent. Style not persistent. Achenes on large fleshy receptacle. **Fragaria.**
As *Fragaria* but achenes on small dry receptacle.
Potentilla.
Epicalyx. Fruit indehiscent with persistent style.
Leaves simple. Scape 1-flowered. Petals 8–9.
Dryas.

Leaves pinnate. Flowers several. Petals 5. **Geum.**

Tribe 2. *Ulmarieae. Flower-axis* flat or slightly concave. *Filaments* narrow at base, soon falling.
Only genus : **Ulmaria.**

Tribe 3. *Sanguisorbeae. Flower-axis* pitcher-shaped, enclosing and usually hardening around two or more achenes.
Epicalyx of bristles. Petals present. Styles terminal.
Agrimonia.
Epicalyx of four leaves. No petals. Styles basal or ventral. **Alchemilla.**
Epicalyx and petals both absent. **Poterium.**

Tribe 4. *Roseae. Flower-axis* tubular, enclosing ∞ carpels, becoming soft when ripe.
Only genus : **Rosa.**

Subfamily *Prunoideae. Carpel* 1, rarely more. *Style* terminal. *Ovules* 2, pendulous. *Fruit* drupaceous.
Only British genus : **Prunus.**

Fam. 5. LEGUMINOSAE. Herbs, shrubs or trees. *Leaves* of spiral stipulate, often compound. *Flowers* usually in racemes, cyclic, heterochlamydeous, 5-merous, hypogynous, usually hermaphrodite and zygomorphic. *Stamens* usually 10, 9 or all often united. *Carpel* mostly solitary with usually ∞ ovules on the ventral suture which is directed backwards. *Style* terminal. *Fruit* a pod or lomentum. *Endosperm* scanty or absent.

Subfamily *Papilionatae. Roots* in symbiosis with *Bacillus radicicola. Flowers* zygomorphic. *Corolla* with descending imbrication in bud.

Tribe 1. *Genisteae.* Shrubs or herbs with simple or palmate leaves. *Leaflets* entire. *Calyx* 2-lipped. All ten *stamens* usually united. *Pod* 2-valved.

Herbs. Leaves palmate. Calyx deeply 2-lipped. Pod coriaceous with spongy septa.　　　***Lupinus.**

Shrubs. Leaves 1-foliate. Calyx with two short deeply toothed lips.　　　**Genista.**

Shrubs. Mature leaves spinescent or scaly. Calyx deeply 2-lipped, coloured.　　　**Ulex.**

Shrubs. Leaves 1–3-foliate or absent. Calyx with two short minutely toothed lips.　　　**Cytisus.**

Tribe 2. *Trifolieae.* Herbs. *Leaves* usually pinnately 3-foliate. *Veins of leaflets* ending in teeth. *Upper stamen* usually free.

A.　All ten stamens united. Keel beaked. **Ononis.**

B.　Upper stamen separate. Keel obtuse.

a.　Corolla deciduous, claws of petals free.

I.　Flowers in heads, umbels, or short racemes.
Pod straight or curved, mostly linear, often beaked.
　　　　　　Trigonella.
Pod usually spiral, sometimes broadly oval and bent.
　　　　　　Medicago.

II.　Flowers in long racemes. Pods obovoid.
　　　　　　Melilotus.

b.　Corolla usually persistent, claws of all or of four lower petals united to the staminal tube.　**Trifolium.**

Tribe 3. *Loteae.* Herbs. *Leaves* pinnate with entire leaflets. *Upper stamen* separate or united. *Pod* 2-valved without longitudinal septum.

Pod enclosed by the calyx.   **Anthyllis.**
Pod long, exserted.   **Lotus.**

Tribe 4. *Astragaleae.* Herbs. *Leaves* pinnate:
leaflets entire. *Upper stamen* usually free. *Pod* 2-valved
with longitudinal septum derived from suture.
Keel obtuse.   Pod-septum from dorsal suture.
                                                      Astragalus.
Keel beaked or tip incurved.   Pod-septum from
ventral suture.   **Oxytropis.**

Tribe 5. *Hedysareae. Upper stamen* usually separate.
Fruit indehiscent, either a lomentum of several 1-seeded
joints or consisting of only one such joint.
Keel obliquely truncate.   Fruit 1–2-seeded, not
jointed.   **⊙Onobrychis.**
Keel beaked.   Fruit jointed, curved, laterally com-
pressed.   **Hippocrepis.**
Keel obtuse.   Fruit terete, many-jointed.
                                                     Ornithopus.
Keel beaked.   Fruit terete or 4-angled, jointed.
                                                     ***Coronilla.**

Tribe 6. *Vicieae.* Mostly herbs. *Leaves* pinnate
with no terminal leaflet but instead a tendril or point.
Cotyledons thick.
Style filiform, hairy below or all round.   **Vicia.**
Style flat, hairy above.   **Lathyrus.**

8. *The flowers have predominantly* 5 *or* 4 *whorls. Apocarpy
and Isomery still occur, but Syncarpy and Oligomery of the Gynaecium preponderate. Pleiomery of Gynaecium rare.* (For γ see
page 54.)

Order 14. **Geraniales.** Herbs (the British species).
Flowers cyclic, heterochlamydeous, usually 5-merous.

Carpels (5–2) often separating from each other when ripe. *Ovules* usually 2–1, more rarely ∞, anatropous, pendulous, with ventral raphe and micropyle directed upwards, or, when more than one ovule is present, single ones sometimes occur with dorsal raphe and micropyle directed downwards.

Suborder 1. GERANIINEAE. *Flowers* heterochlamydeous, mostly actinomorphic. Usually obdiplostemonous (i.e. *stamens* twice as many as petals, and *carpels* when isomerous opposite petals). Haplostemony more rarely, or individual stamens may abort in zygomorphic flowers. *Anthers* opening by longitudinal slits. *Carpels* isomerous or oligomerous. *Ovules* with two integuments.

Fam. 1. GERANIACEAE. Herbs with lobed or divided, often stipulate *leaves*. *Flowers* 5-merous, actinomorphic. No real disc present. *Petals* imbricate or convolute in bud. *Stamens* 10, sometimes only five fertile. *Carpels* usually 5, each with 1–2 ovules. *Capsule* splitting into five beaked mericarps. *Seeds* with endosperm.

Tribe 1. *Geranieae*. *Mericarps* separating by means of elastic awns.
Fertile stamens 10. Styles glabrous within, twisting in flat spiral after dehiscence. **Geranium†**.
Fertile stamens 5. Styles silky within, twisting in helicoid spiral after dehiscence. **Erodium.**

† The so-called *Geraniums* of cultivation belong to the South African genus *Pelargonium*, which differs from the other *Geranieae* in having a sepaline spur which is adnate to the pedicel and best seen in section. The flowers are slightly zygomorphic. The fruit resembles that of *Erodium*.

Fam. 2. OXALIDACEAE. Perennial herbs. *Leaves* often ternate, showing sleep-movements. *Flowers* 5-merous. No real *disc. Stamens* 10, united below. *Carpels* isomerous. *Fruit* a capsule or berry, loculicidal if dehiscent. *Endosperm* fleshy.

Only British genus: **Oxalis.**

Fam. 3. LINACEAE. Herbs with simple, entire *leaves. Flowers* 4-5-merous, actinomorphic. *Petals* often convolute in bud. No real *disc. Stamens* 5-20, united in a ring below. *Gynaecium* quite syncarpous. *Ovary* often with extra, imperfect dissepiments from dorsal sutures. *Fruit* a capsule or drupe.

Tribe *Eulineae. Stamens* in one whorl. *Fruit* a capsule. Flowers 5-merous. Sepals entire. **Linum.** Flowers 4-merous. Sepals with 2-3 teeth. **Radiola.**

Suborder 2. POLYGALINEAE. *Flowers* actinomorphic or zygomorphic. *Stamens* in two whorls. *Anthers* opening by pores. *Carpels* (2), median.

Fam. 4. POLYGALACEAE. Herbs or shrubs. *Leaves* usually simple, entire, exstipulate. *Flowers* 5-merous, zygomorphic. *Sepals* 5, of which two are petaloid and wing-like. *Petals* by abortion only 3. *Stamens* 8. *Carpels* with one rarely with 2-4 ovules. *Fruit* a capsule or drupe.

Only British genus: **Polygala.**

Suborder 3. TRICOCCAE. *Flowers* actinomorphic, diclinous, often very reduced. *Carpels* usually (3). *Seed* mostly with caruncle.

Fam. 5. EUPHORBIACEAE. Herbs, shrubs or trees, often with latex. *Leaves* usually spiral, often stipulate. *Inflorescence* mostly compound. *Stamens* as many as sepals or twice as many or ∞ or 1. *Carpels* usually (3). *Fruit* usually a capsule splitting into three mericarps. *Endosperm* copious.

Subfamily *Crotonoideae.* Each *carpel* with one *ovule.*

Tribe 1. *Acalypheae. Latex* absent. *Calyx* valvate. *Flowers* racemose. ♂ usually without corolla.

Only British genus: **Mercurialis.**

Tribe 2. *Euphorbieae. Latex* present. *Flowers* in *cyathia*†. ♂ without corolla and usually without calyx, and with only one stamen.

Only British genus: **Euphorbia.**

Suborder 4. CALLITRICHINEAE. Characters of the family. (Systematic position doubtful. Possible affinities with Verbenaceae.)

Fam. 6. CALLITRICHACEAE. Slender glabrous herbs, often submerged. *Leaves* opposite entire, upper often rosulate and floating. *Flowers* monoecious, naked, often with two falcate bracteoles. ♂ of one terminal *stamen.* ♀ of two transverse *carpels.* A longitudinal septum makes ovary 4-loc. *Fruit* of four drupaceous mericarps. *Seeds* with endosperm.

Only genus: **Callitriche.**

Order 15. **Sapindales.** Mostly woody plants. (Impatiens is the only British herbaceous genus.) Characters of Geraniales but ovules in the reverse position, either pendulous with dorsal raphe and micropyle directed upwards, or ascending with ventral raphe and micropyle directed downwards.

† A *Cyathium* is a condensed cymose inflorescence resembling a single flower.

Suborder 1. BUXINEAE. *Flowers* haplochlamydeous. *Ovules* with two integuments.

Fam. 1. BUXACEAE. Woody plants. *Leaves* entire, evergreen, exstipulate. *Flowers* diclinous, sometimes with rudiments of the aborted sporophylls, actinomorphic. *Stamens* 4–∞. *Carpels* usually (3) each with 2–1 ovules. *Styles* separate. *Fruit* a loculicidal capsule or drupe. *Seeds* with endosperm.

Only British genus (leaves opposite): **Buxus.**

Suborder 2. EMPETRINEAE. *Flowers* heterochlamydeous. Each *carpel* with one ascending ovule with one integument. *Carpels* not separating when ripe.

Fam. 2. EMPETRACEAE. Ericoid shrubs with linear exstipulate leaves deeply furrowed beneath. *Flowers* diclinous with rudiments of the aborted sporophylls, actinomorphic. K, C and A 2–3. G (2–9). *Fruit* a drupe.

Only British genus (leaves spiral, stamens 3): **Empetrum.**

Suborder 3. CELASTRINEAE. *Flowers* heterochlamydeous, always actinomorphic, diplostemonous or haplostemonous. *Gynaecium* tending towards oligomery, rarely isomerous.

Fam. 3. AQUIFOLIACEAE. Trees with spiral, mostly evergreen, simple leaves. *Flowers* 4–5-merous, dioecious, actinomorphic. *Disc* absent. *Petals* often connate at base and adnate to the isomerous stamens. *Carpels* (4–6) each with 1–2 pendulous ovules with one integument. *Fruit* a drupe.

Only British genus (petals connate at base): **Ilex.**

Fam. 4. CELASTRACEAE. Trees or shrubs with simple opposite or spiral leaves. *Stipules* absent or

C.   5

deciduous. *Flowers* small, often green, 4–5-merous, mostly hermaphrodite. *Disc* usually conspicuous. *Petals* imbricate. *Stamens* 4–5, inserted on the disc. *Carpels* (2–5) each with ∞–1 ovules. *Fruit* a capsule or berry. *Seeds* often arillate.

Only British genus (leaves opposite, carpels isomerous):   **Euonymus.**

*Fam. 5. STAPHYLEACEAE. Shrubs or trees. *Leaves* usually opposite and pinnate, often with stipules and stipels. *Flowers* in panicles, 5-merous. *Stamens* 5, inserted outside the disc. *Carpels* (2–3), free above. *Ovules* ∞–few. *Fruit* capsular. *Embryo* straight. *Endosperm* copious, fleshy.

Capsule inflated.   ***Staphylea.**

Suborder 4. SAPINDINEAE. *Flowers* heterochlamydeous, typically diplostemonous but some stamens and carpels abort, actinomorphic or obliquely zygomorphic. *Ovules* with two integuments.

Fam. 6. ACERACEAE. Trees. *Leaves* opposite, simple, lobed or pinnate, exstipulate. *Flowers* actinomorphic, axis flat or concave. K and C 4–10. A 4–10, usually 8. G (2) each with two almost orthotropous ovules. *Fruit* indehiscent, winged, usually 1-seeded. No *endosperm*.

Only British genus (fruit wings lateral):   **Acer.**

*Fam. 7. HIPPOCASTANACEAE. Trees. *Leaves* opposite, palmate, exstipulate. *Flowers* in conspicuous, primarily racemose inflorescences, obliquely zygomorphic. K 5. C 4–5. A 5–8. G (3) each with two ovules. *Capsule* 1-seeded, 2–3-valved. No *endosperm*.

Only British genus (calyx tube long):   ***Aesculus.**

Suborder 5. BALSAMININEAE. *Flowers* heterochlamydeous, zygomorphic, haplostemonous, with united anthers.

Fam. 8. BALSAMINACEAE. Herbs or shrubs. *Leaves* opposite or alternate. *Flowers* always hermaphrodite. *Sepals* usually 3 (2 anterior, if present, minute), posterior spurred. *Petals* primitively 5, *anterior* large, external in bud, *lateral* and *posterior* connate on each side so that there are apparently only three petals. *Stamens* 5, filaments short, anthers cohering. *Carpels* (5) each with ∞ ovules. *Capsule* loculicidal, opening elastically, rarely indehiscent. *Seeds* ∞. *Endosperm* absent.

Only British genus (capsule opening elastically):
Impatiens.

Order 16. **Rhamnales.** *Flowers* cyclic, diplochlamydeous, sometimes apopetalous. *Stamens* in one whorl opposite the petals. *Carpels* (5–2) each with 1–2 ascending ovules, with dorsal, lateral or ventral raphe and two integuments.

Fam. RHAMNACEAE. *Stems* usually woody, rarely herbaceous, often climbing. *Leaves* simple, often 3–5-nerved. *Stipules* small. *Flowers* small, greenish or yellowish, often in axillary cymes, 5–4-merous, perigynous or epigynous. *Calyx lobes* triangular, valvate. *Petals* small, often clawed and hooded and enclosing the stamens, or absent. *Stamens* inserted on calyx tube at edge of disc. *Carpels* (5–2). *Fruit* various, often a drupe. *Seeds* mostly with endosperm.

Tribe *Rhamneae*. *Serial buds* not present (as in *Colletia*). *Ovary* superior or inferior.

Only British genus (disc lining the calyx tube):
Rhamnus.

Order 17. **Malvales.** *Flowers* cyclic (*androecium* not always so), heterochlamydeous, hermaphrodite, rarely

5—2

zygomorphic. *Calyx* and *corolla* usually 5-merous, *sepals* mostly valvate in bud. *Stamens* ∞ or in two whorls, the inner of which is branched. *Carpels* (2–∞) each with 1–∞ anatropous ovules with two integuments. Suborder MALVINEAE. *Sepals* mostly valvate: *mucilage* present.

Fam. 1. TILIACEAE. Mostly trees with entire or toothed stipulate *leaves. Mucous-tubes* present in pith and bark. *Flowers* usually hermaphrodite. K and C 5-merous. *Calyx* valvate. *Corolla* sometimes absent. *Stamens* usually ∞ inserted on disc free or in 5–10 bundles. *Anthers* with two thecae. *Carpels* (2–∞) each with 1–∞ ovules. *Style* 1. *Fruit* 2–∞-locular or 1-locular by abortion. *Endosperm* mostly present.

Only British genus (wing-like bract adnate to peduncle): **Tilia.**

Fam. 2. MALVACEAE. Herbs, shrubs or trees with simple or lobed stipulate *leaves. Sepals* 5, valvate, often with *epicalyx. Petals* 5, adnate to base of staminal column, twisted in bud. *Stamens* usually ∞, *filaments* forming a tube. *Anthers* with one theca. *Pollen grains* large, spiny. *Carpels* (5–∞) each with 1–∞ ovules. *Styles* as many or twice as many as *carpels. Fruit* a capsule or schizocarp. *Seeds* often woolly. *Endosperm* scanty or absent.

Tribe *Malveae. Style-arms* as many as carpels. *Fruit* a schizocarp.

Epicalyx of 3–6 connate segments. Axis longer than fruits. **Lavatera.**

Epicalyx of 6–9 connate segments. Axis shorter than fruits. **Althaea.**

Epicalyx of 3 distinct segments. **Malva.**

ε. *Flowers spirocyclic or of* 5-4 *whorls. Syncarpy the rule* (*Apocarpy in the more primitive, exotic families*). *Tendency for gynaecium to become sunk in flower-axis.* (For ♂ see page 61, for ♀ see page 71.)

Order 18. **Parietales.** *Flowers* spirocyclic or cyclic often with indefinite number of *stamens* and *carpels*, heterochlamydeous, rarely apopetalous, hypogynous or epigynous. *Carpels* more or less united often with parietal *placentae*, which, however, may meet in the middle. *Ovules* very seldom basal.

The suborders of Parietales may stand in phylogenetic relationship to some of the earlier orders, particularly Ranales and Rhoeadales. Suborder Flacourtiineae (see page 71) shows possible affinities with Cucurbitaceae. Yet the Cucurbitaceae possess so many important peculiarities that they cannot have been derived directly from the Flacourtiineae.

Suborder 1. THEINEAE†. *Stamens* often ∞. *Gynaecium* free, on convex or flat flower-axis. *Placentation* often axile. *Endosperm* containing oil and proteid grains.

Fam. 1. GUTTIFERAE. Trees or shrubs, more rarely herbs (most British species are herbaceous). *Leaves* opposite, entire, usually exstipulate, often with pellucid glands. K and C often 5, *sepals* imbricate.. *Stamens* often ∞ and united in bundles. *Carpels* (3–5) with ∞ –1 ovules with two integuments. *Placentation* parietal or axile. *Seeds* without endosperm.

Subfamily *Hypericoideae. Stamens* ∞ in 2–5 bundles. *Styles* usually free. *Fruit* a capsule, berry or drupe.

Tribe *Hypericeae.* Herbs and shrubs. *Ovary* often 3–5-loc. *Capsule* septicidal.

Only British genus (petals 5, unequal-sided) :

Hypericum.

† Cf. *Guttiferales* of Bentham and Hooker.

Suborder 2. TAMARICINEAE. *Stamens* in whorls or if ∞ in bundles. *Gynaecium* free on flat flower-axis. *Endosperm* starchy or absent.

Fam. 2. ELATINACEAE. Usually small plants growing in wet places and rooting at the nodes. *Leaves* opposite or whorled, stipulate. *Flowers* small, cyclic, 2–5-merous. *Sepals* and *petals* each 2–5, imbricate. *Stamens* in one or two whorls. *Ovary* 2–5-loc with ∞ axile ovules with two integuments. *Styles* as many as loculi, free. *Capsule* septicidal.

Only British genus (leaves spathulate):   **Elatine.**

Fam. 3. FRANKENIACEAE. Usually small shrubs. *Stems* jointed at the nodes. *Leaves* opposite, exstipulate. *Flowers* 4–6-merous. *Calyx* gamosepalous. *Petals* free, with ligular appendages. *Stamens* 4–∞. *Carpels* 4–2 with ∞ ascending ovules on parietal placentae. *Style* slender. *Stigma* 2–5-lobed. *Capsule* dehiscing between placentae. *Seeds* with endosperm.

Only British genus (petals 5, usually with appendages):   **Frankenia.**

*Fam. 4. TAMARICACEAE. Herbs or shrubs with minute alternate, entire leaves. *Flowers* mostly 4–5-merous. *Stamens* as many as petals or twice as many or ∞ in groups. *Carpels* (5–2). *Ovary* usually 1-loc with ∞ basal or parietal ovules. *Fruit* a capsule. *Seeds* hairy.

Tribe *Tamariceae*. *Flowers* racemose. *Seeds* with apical tuft of hairs.

Only British genus (stamens free):   ***Tamarix.**

Suborder 3. CISTINEAE. *Stamens* ∞, not in bundles. *Gynaecium* free on flat or convex axis. *Endosperm* with starch.

Fam. 5. CISTACEAE. Herbs and shrubs. *Leaves* mostly opposite. Stellate and glandular hairs, the latter with essential oils, often present. K 5–3. C 5–3 or o. A ∞. G (5–10). *Ovary* usually 1-celled with parietal placentae with ∞ or 2 more or less orthotropous ovules. *Capsule* dehiscing between placentae. *Seeds* with endosperm. Only British genus (ovules ∞, capsule 3-valved):
Helianthemum.

Suborder 4. FLACOURTIINEAE. *Stamens* often 5. *Gynaecium* free on convex or tubular axis, rarely adnate to axis. *Endosperm* copious containing oil and proteid bodies.

Fam. 6. VIOLACEAE. Herbs or shrubs with alternate stipulate leaves. *Flowers* 5-merous (except gynaecium), actinomorphic or zygomorphic. *Petals* sometimes united. *Stamens* 5. *Carpels* (3) each with 1–∞ anatropous ovules with two integuments on parietal placentae. *Fruit* a loculicidal capsule or berry.

Tribe *Violeae. Corolla* zygomorphic, *two anterior stamens* spurred.

Only British genus (flowers solitary in the leaf axils):
Viola.

ζ. *The flowers are cyclic and the sinking of the gynaecium into the hollow flower-axis is general: connation of gynaecium and flower-axis predominates.* (For ε see page 69.)

Order 19. **Myrtiflorae.** Herbs, shrubs, or trees, bundles often bicollateral. *Flowers* cyclic, heterochlamydeous, rarely apopetalous, haplo- or diplostemonous, usually actinomorphic. *Flower-axis* concave. *Gynaecium* syncarpous and mostly united to axis.

Suborder 1. THYMELAEINEAE. Usually shrubs with entire leaves. *Flower-axis* more or less tubular (except in ♂ Elaeagnaceae). *Gynaecium* of 2–4 carpels, free from the flower-axis.

Fam. I. THYMELAEACEAE. Shrubs or trees with acrid juice and tenacious, reticulate bast. *Leaves* entire, alternate or opposite, exstipulate. *Flowers* hermaphrodite, heterochlamydeous or apopetalous, diplo- or haplostemonous, 5–4-merous. G usually I with one pendulous ovule. *Style* I, terminal or lateral. *Fruit* often a drupe.

Subfamily *Thymelaeoideae*. *Flowers* diplostemonous or haplostemonous. *Petals* scale-like or absent. *Carpel* I with one ovule.

Only British genus (filaments and style very short) :
Daphne.

Fam. 2. ELAEAGNACEAE. Shrubs or trees with silvery or brown scales. *Flower-axis* in ♂ flowers flat, in ♀ and ⚥ flowers tubular. *Flowers* mostly 4-merous, homochlamydeous. *Androecium* diplostemonous in ♂ flowers ; in ⚥ haplostemonous and with stamens opposite perianth lobes. *Carpel* I with one basal ovule. *Fruit* a nut enclosed in the fleshy axis.

Only British genus (sepals 2, stamens 4): **Hippophaë.**

Suborder 2. MYRTINEAE. Herbs or shrubs. *Leaves* more often opposite. *Flowers* with tubular axis and 2–∞ carpels forming syncarpous gynaecium which is usually united to axis. *Ovules* with one integument.

Fam. 3. LYTHRACEAE. Herbs or shrubs. *Leaves* opposite or whorled, entire. *Flowers* heterochlamydeous or apopetalous, usually 4–6-merous, hermaphrodite, actinomorphic or zygomorphic. *Flower-axis* cup-shaped or tubular. *Calyx* valvate with intersepalar stipules. *Petals* on edge of flower-axis, sometimes absent. *Stamens* twice as many as sepals or 1–∞, inserted deeper than the petals. *Carpels* (2–6) each with ∞–2 ovules, free from the *flower-axis. Ovary* 2–6-loc, *ovules* ∞, axile. *Style* 1 with capitate stigma. *Fruit* a capsule. *Seeds* without endosperm.

Tribe *Lythreae. Septa of ovary* imperfect upwards. Flower-axis tubular. Corolla large. Stamens 12. Fruit dehiscent. **Lythrum.** Flower-axis campanulate. Corolla small or absent. Stamens 6. Fruit indehiscent. **Peplis.**

Fam. 4. EPILOBIACEAE (Onagraceae). Usually herbs. *Leaves* alternate or opposite, exstipulate. *Flowers* heterochlamydeous, usually hermaphrodite and actinomorphic. *Flower-axis* tubular. K 2–4, P 2–4, rarely absent. A 4–8, sometimes partially staminodal. G usually (4), united to axis, each with ∞–1 ovules. *Style* 1. *Stigma* entire or 4-lobed. *Fruit* various, usually with many seeds. *Endosperm* scanty or absent.

Tribe *Oenothereae. Ovary* quite inferior. Calyx persistent. Petals short or absent. Stamens 4. Capsule short. **Ludwigia.** Calyx deciduous. Petals 4, pink or purple. Stamens 8. Capsule long. Seeds with hairs. **Epilobium.** Calyx deciduous. Petals 4 (yellow in Brit. sp.). Capsule long. Seeds with membranous margin. ***Oenothera.** Calyx deciduous. Petals 2, white. Stamens 2. Fruit dry, indehiscent. **Circaea.** Shrubs. Flowers pendulous. Calyx petaloid. Fruit a berry. ***Fuchsia.**

Fam. 5. HALORRHAGACEAE. Herbs of very diverse
habit. *Flowers* small, heterochlamydeous, or often apopetalous. *Stamens* twice as many as *sepals* or fewer.
Anthers long, 4-angled. *Carpels* usually (4) united to
the tubular axis, each with one pendulous ovule. *Styles*
separate. *Fruit* a nut or drupe. *Seeds* with endosperm.

Subfamily 1. *Halorrhagoideae. Petals* 2–4 or 0.
Carpels (2–4) each with one ovule.
Only British genus (leaves whorled, deeply pin-
natifid. Petals 2–4):   **Myriophyllum.**

Suborder 3. HIPPURIDINEAE. *Flowers* epigynous.
Stamen 1. *One carpel* with one ovule without integument.

Fam. 6. HIPPURIDACEAE. Aquatic herbs. *Leaves*
linear, in whorls. *Flowers* minute, axillary, naked.
Carpel 1 with one undivided style, stigmatic throughout
its length.
Only genus (stems simple, erect):   **Hippuris.**

Order 20. **Umbellales (Umbelliflorae).** *Flowers*
mostly in umbels, cyclic, heterochlamydeous, usually
haplostemonous, epigynous, 4–5-merous. *Carpels* (5–1),
each carpel with one (rarely two) pendulous anatropous
ovules with one integument. *Seeds* with copious endosperm.

Fam. 1. ARALIACEAE. Mostly trees and shrubs,
often twining or climbing by aid of roots. *Leaves* alternate, simple or compound, often stipulate. *Flowers*
mostly in umbels or heads, 5-merous. *Calyx* often with
obscure limb. *Stamens* usually 5. *Carpels* (1–∞) each

usually with one ovule. *Micropyle* facing outwards. *Fruit* a berry or drupe.

Tribe *Schefflereae*. *Petals* valvate. *Endosperm* sometimes lobulate.

Only British genus (leaves simple, endosperm lobulate): **Hedera.**

Fam. 2. UMBELLIFERAE. Herbs with hollow internodes. *Leaves* alternate with conspicuous *sheath*, and usually compound *lamina*. *Flowers* in simple or compound umbels, 5-merous, haplostemonous. *Calyx* usually inconspicuous. *Petals* epigynous, tips often inflexed. *Stamens* at base of stylopodium. *Anthers* versatile. G (2), median. *Styles* 2 arising from epigynous disc (*stylopodium*). *Fruit* separating into two *mericarps* pendulous on the *carpophore*, each *mericarp* with five *primary ridges*, i.e. two lateral next to the commissure and three dorsal. Four *secondary ridges* are sometimes present between the primary. Oil canals (*vittae*) occur in the grooves between the *primary ridges*: two or more are present on the commissural face of the mericarp. *Seed* adherent to pericarp. *Endosperm* copious. *Embryo* minute.

Subfamily 1. *Hydrocotyloideae*. *Leaves* often simple (peltate in the only British sp.). *Flowers* in *heads* or simple umbels. *Fruit* with woody fibrous *endocarp*. *Carpophore* undivided. *Vittae* absent or in the primary ridges: never in the furrows.

Tribe 1. *Hydrocotyleae*. *Fruit* laterally compressed. (*Commissure* narrow.)
Only British genus: **Hydrocotyle.**

Subfamily 2. *Saniculoideae. Leaves* various. *In-florescence* as above. *Endocarp* soft. *Styles* long sur-rounded at base by annular *stylopodium. Stigmas* capitate. *Vittae* various.

Tribe 2. *Saniculeae. Commissure* broad. Leaves palmately divided. Umbels irregularly com-pound. Fruit globose, with hooked bristles. Ridges obscure. **Sanicula.** Leaves palmate. Umbels simple. Bracts large, coloured. Fruit ovoid. Ridges wrinkled or toothed. ***Astrantia.** Leaves with spinous teeth. Flowers capitate. Fruit rough, ridges absent. **Eryngium.**

Subfamily 3. *Apioideae. Leaves* various. *Flowers* usually in compound umbels. *Endocarp* soft (there may be layer of *stereom* beneath *epidermis*). *Style* arising from apex of stylopodium. *Vittae* at first in the furrows, later variously arranged.

A. *Haplozygieae. Primary ridges* more con-spicuous than the secondary. *Vittae* usually obvious in the furrows.

α. *Seed* furrowed ventrally by the *raphe.*

Tribe 3. *Scandicineae. Crystal-glands* present around *carpophore.*

Subtribe 1. *Scandicinae. Fruit* long-cylindrical and beaked, smooth or with short spines.
i. Vittae several in each furrow. **Conopodium.**
ii. Vittae 1 in each furrow, or absent.
Fruit smooth, over 2 cm. long, with prominent obtuse ridges and long beak. **Scandix.**

Umbelliferae 77

Fruit slightly rough, over 2 cm. long. Ridges very
acute. Carpophore split to middle. **Myrrhis.**
Fruit rather rough, 5 mm. long. Ridges vanishing
upwards. Carpophore undivided or shortly bifid.
Chaerophyllum.
Fruit smooth, under 1 cm. long. Beak short, ribbed.
Anthriscus.
Subtribe 2. *Caucalinae. Fruit* ovoid, *secondary ridges*
spinous.
Only British genus: **Caucalis** (including **Torilis**).

Tribe 4. Coriandreae. Crystal-glands absent around
carpophore. Fruit with woody layers beneath the
epidermis. Mericarps coherent.
Coriandrum.

Tribe 5. *Smyrnieae.* No *crystal-glands. Fruit*
ovoid. *Commissure* narrow. *Mericarps* separating.
One vitta in each furrow; ridges entire, slender.
Physospermum.
Several vittae in each furrow. Stylopodium flattened.
Ridges crenate. **Conium.**
Several vittae in each furrow. Stylopodium conical.
Ridges obscure. **Smyrnium.**

β. *Seed* flat ventrally. *Raphe* often projecting
towards carpophore.

Tribe 6. *Ammineae. Dorsal and commissural primary
ridges* all alike. *Seed* semicircular in section.

Subtribe 1. *Carinae. Ridges* not prominent. *Commissure* narrow.

1. Petals entire, with acute or inflexed tips. Vittae
1–3 in each furrow.

Leaves simple, entire. Flowers yellow. **Bupleurum.**
Leaves compound. Flowers white, dioecious. ♂
flowers with narrower petals. **Trinia.**
Leaves compound. Flowers white, hermaphrodite.
Apium.

2. Petals 2-lobed. Tip long, inflexed. One vitta
in each furrow.
Calyx teeth minute or absent. Vittae as long as
fruit. **Carum.**
Calyx teeth minute or absent. Vittae only in upper
half of fruit. **Sison.**
Calyx teeth leafy, ovate, acute. Vittae long. **Cicuta.**

3. Petals as 1, but vittae usually several in each
furrow.
Leaves pinnate. Calyx teeth acute. Ridges obtuse.
Sium.
Leaves twice 3-nate. Calyx teeth absent. Ridges
slender. Vittae o. **Aegopodium.**
Calyx teeth absent. Vittae many. **Pimpinella.**

Subtribe 2. *Seselinae.* *Ridges* prominent, sometimes
winged, lateral ones forming continuation of commissural
face of fruit.

1. Fruit almost cylindrical. Ridges not thickened
or corky.
Calyx teeth minute. Petals white, notched. **Seseli.**
Calyx teeth o. Petals yellow, entire. **Foeniculum.**

2. Fruit as above. Primary ridges acute; outer coat
of pericarp loose, corky. **Crithmum.**

3. Fruit as above. Primary ridges thick, lateral
ridges forming a corky rim round the carpel.

Bracteoles short, whorled. Oenanthe.
Bracteoles long, unilateral. Aethusa.
4. Fruit as above, lateral ridges thickened or winged. Silaus.
5. Fruit dorsally compressed. Primary ridges broad, thick.
Seed grooved ventrally, vittae several. Meum.
Seed flat or slightly concave ventrally, vittae many or obscure. Ligusticum.
Seed biconvex: one vitta in each of the dorsal furrows. Selinum.

Tribe 7. *Peucedaneae.* Lateral *ridges* winged, much broader than the three dorsal ridges which are often only feebly developed. Seed narrow in section.

Subtribe 1. *Angelicinae.* Lateral *ridges* of opposite carpels not appressed but gaping.
Only British genus:
Angelica (including *Archangelica).

Subtribe 2. *Ferulinae.* Lateral *ridges* appressed, forming a wing round the fruit.
Only British genus: Peucedanum.

Subtribe 3. *Tordyliinae.* As *Ferulinae* but wing hardened.
Vittae clavate. Margins of wing thin. Heracleum.
Vittae slender. Margins of wing thick. Tordylium.

B. *Diplozygieae.* Secondary *ridges* as large as or larger than the primary.

Tribe 8. *Dauceae.* Secondary *ridges* spinous.
Daucus.

Fam. 3. CORNACEAE. Trees or shrubs. *Leaves* opposite or spiral, usually entire, exstipulate. *Flowers* small, in cymes, umbels, or heads, 4–5-merous, mostly haplostemonous. *Carpels* (4–1) with an epigynous disc, each carpel with one ovule. *Micropyle* facing outwards or inwards. *Style* 1. *Fruit* a drupe or berry.

Subfamily *Cornoideae*. *Ovary* inferior. *Raphe* dorsal.

Only British genus (leaves opposite. Petals 4, valvate) :   **Cornus.**

Subclass 2. **METACHLAMYDEAE** (Sympetalae). *Perianth* in advanced stage of development, always originally double, and the inner whorl gamopetalous. Polypetalous forms occur but they are closely related to gamopetalous forms.

A. **Polypetaly, as well as sympetaly, occurs. Two whorls or one whorl of stamens. Hypogyny predominates, but epigyny also occurs.** (For B see page 85.)

Order 1. **Ericales.** Shrubs and herbs, rarely trees, with simple *leaves*. *Flowers* 4–5-merous, obdiplostemonous, or the antipetalous whorl of *stamens* not developed. *Petals* free or united. *Stamens* hypogynous or epigynous, more rarely united at base with the *petals*. *Carpels* (2–∞), when isomerous opposite petals. *Ovary* superior or inferior. *Ovules* with one *integument*.

Fam. 1. PIROLACEAE. Evergreen or colourless perennial herbs with spiral *leaves*. *Flowers* solitary or racemose, 5–4-merous, obdiplostemonous. *Petals* free or united. *Stamens* hypogynous. *Carpels* (5–4) each with ∞ minute *ovules* on fleshy *placentae*. *Capsule*

loculicidal. *Seeds* with loose *testa* and fleshy *endosperm.*
Embryo of few cells without cotyledons.

Subfamily 1. *Piroloideae.* *Anthers* recurved before
flowering, then erect, opening by two *pores.* *Pollen* in
tetrads.

Only British genus (herbs with broad evergreen
leaves): Pirola (including Moneses).

Subfamily 2. *Monotropoideae.* *Anthers* always
erect at the apex of the filament, opening by slits.
Pollen simple.

Tribe *Monotropeae.* *Ovary* 4-5-loc below, 1-loc
above.
Only British genus (colourless root-parasites):
Monotropa.

Fam. 2. ERICACEAE. Usually low shrubs. *Leaves*
mostly evergreen. *Flowers* solitary or racemose, 5-4-
merous, obdiplostemonous. *Corolla* usually sympetalous.
Stamens on the edge of a hypogynous or epigynous disc.
Anther-thecae free, tubular, often spreading above and
with basal or dorsal *appendages.* *Pollen* in tetrads.
Gynaecium syncarpous. *Ovules* 1-∞ in each carpel on
axile *placentae.* *Style* 1, *stigma* capitate. *Fruit* various.
Seeds with loose *testa* and copious *endosperm.* *Embryo*
often very short.

A. *Septicidal capsule.* *Petals* free or united. *Anthers*
without appendages. *Testa* loose, often winged.

Subfamily 1. *Rhododendroideae.* As above. *Buds*
scaly.

c. 6

*Tribe 1. *Ledeae.* *Flowers* actinomorphic. *Petals* free. *Seeds* with long wings.

(Petals 5, spreading.)   *Ledum.

*Tribe 2. *Rhododendreae.* *Corolla* sympetalous, often weakly zygomorphic. *Seeds* surrounded by broad wing. (Stamens exserted.)

*Rhododendron (including *Azalea).

Tribe 3. *Phyllodoceae.* *Petals* usually united. *Seeds* globose or 3-angled, not winged. *Embryo* cylindrical. Corolla campanulate. Stamens 5. Anthers opening by slits. Capsule 2–3-valved.   Loiseluria. Corolla urceolate. Stamens 10. Anthers opening by pores. Capsule 5-valved.   Phyllodoce. Corolla urceolate. Stamens 8. Anthers opening by pores. Capsule 4-valved.   Dabeocia.

B. *Berry, drupe* or *loculicidal capsule.* *Petals* united, deciduous. *Anthers* appendaged or prolonged into tubes.

Subfamily 2. *Arbutoideae.* *Ovary* superior.

Tribe 4. *Andromedeae.* *Buds* scaly. *Capsule* loculicidal.

Only British genus (Corolla pitcher-shaped):
                                        Andromeda.

Tribe 5. *Arbuteae.* *Buds* naked. *Berry* or *drupe.* *Anthers* with two long reflexed appendages.

Berry many-seeded with firm endocarp.   Arbutus. Drupe with several 1-seeded stones. Arctostaphylos.

Subfamily 3. *Vaccinioideae.* *Ovary* inferior.

Tribe 6. *Vaccinieae.* *Petals* united. *Stamens* epigynous. *Ovary* sharply demarcated from peduncle.

Only British genus (Filaments straight, Fruit a berry):
                    Vaccinium (including Oxycoccos).

ERRATUM

page 83, line 9, *for* sepaloid *read* petaloid.

C. *Capsule. Petals* united, persistent, scarious. *Anther-thecae* opening by pores above. *Appendage* usually on back of connective.

Subfamily 4. *Ericoideae.*

Tribe 7. *Ericeae. Buds* naked. *Leaves small. Carpels* with many *ovules. Capsule* many-seeded. Calyx small, green. Corolla with four lobes. Capsule loculicidal. **Erica.** Calyx sepaloid, exceeding corolla. Corolla deeply divided. Capsule septicidal. **Calluna.**

Order 2. **Primulales.** *Flowers* 5 (rarely 4–8)-merous, usually actinomorphic. *Corolla* nearly always gamopetalous. *Stamens* in one whorl, inserted on corolla opposite the lobes (*antipetalous*). *Ovary* 1-loc with ∞–1 ovules which have two *integuments* on free central or basal *placenta.*

Fam. PRIMULACEAE. Mostly perennial herbs. *Leaves* usually spiral, exstipulate. *Stamens* opposite corolla lobes. *Ovary* usually superior with ∞ whorled or spirally arranged ovules on a free central *placenta. Style* 1. *Capsule* with usually many-angled and facetted *seeds* sunk in cavities on the *placenta.*

Tribe 1. *Androsaceae. Corolla* imbricate or quincuncial in bud, *lobes* eventually erect or spreading, never reflexed (in British genera).

Subtribe 1. *Primulinae.* Land plants. Corolla lobes entire or bifid. Capsule with valves.
Only British genus (Leaves broad. Flowers conspicuous. Corolla-tube long): **Primula.**

Subtribe 2. *Hottoniinae.* Water plants. Corolla lobes entire. Capsule with valves.
Only genus (Leaves all submerged, pinnately divided):
Hottonia.

*Tribe 2. *Cyclamineae. Rootstock* often tuberous. *Corolla lobes* reflexed.
Only British genus (Scapes 1-flowered): ***Cyclamen.**

Tribe 3. *Lysimachieae. Corolla tube* short. *Lobes* twisted in bud, never reflexed.

Subtribe 1. *Lysimachiinae.* Capsule with valves.
Leaves subrosulate towards top of scape. Corolla white, lobes 5–9. **Trientalis.**
Leaves spiral, opposite or whorled. Corolla yellow, lobes 5. **Lysimachia.**

Subtribe 2. *Anagallidinae.* Capsule opening by lid.
Corolla absent. Calyx campanulate, coloured. **Glaux.**
Corolla shorter than calyx. Filaments glabrous.
Centunculus.
Corolla longer than calyx. Filaments villous.
Anagallis.

Tribe 4. *Samoleae. Corolla* quincuncial in bud. *Ovary* half inferior.
Only genus (Bracts displaced on to the pedicels):
Samolus.

Order 3. **Plumbaginales.** *Corolla* polypetalous or gamopetalous. *Stamens* in one whorl opposite the petals. *Gynaecium* (5). *Styles* 5 or style 5-fid. *Ovary* 1-loc with one basal anatropous *ovule* with long *funicle* and two *integuments.* (Perhaps connected with *Centrospermae.*)

Fam. PLUMBAGINACEAE. Often maritime herbs, rarely shrubs. *Leaves* entire, spiral, often radical. *Inflorescence* mostly compound. *Flowers* with bracts and bracteoles. *Calyx* often scarious and coloured. *Petals* free or united. *Fruit* enclosed in calyx.

Tribe *Staticeae. Inflorescence* compounded of cincinni. *Stamens* united to the *corolla. Styles* united only at base.

Cymes very dense. Styles hairy.	**Armeria.**
Cymes lax. Styles glabrous.	**Statice.**

B. **Sympetaly dominant. Stamens always in one whorl. Union of Carpels sometimes imperfect. Hypogyny usual.** (For A see p. 80, for C see p. 87.)

Order 4. **Contortae.** Shrubs and herbs, rarely trees. *Leaves* usually opposite, entire and exstipulate. *Flowers* mostly 5-merous, more rarely 2–6-merous, mostly sympetalous. *Corolla* usually twisted in bud. *Stamens* usually isomerous (two in *Oleaceae*) and inserted on the *corolla*, rarely hypogynous. *Carpels* 2, sometimes free below.

Suborder 1. OLEINEAE. *Stamens* 2. *Ovules* with one integument.

Fam. 1. OLEACEAE. Woody plants, sometimes climbing, rarely herbs. *Leaves* opposite or whorled, simple or pinnate. *Flowers* racemose or cymose, 2–6-merous, mostly sympetalous, more rarely with free petals or apopetalous, hermaphrodite or diclinous. *Corolla* 4, 5, or 6, free or united. *Stamens* 2, *filaments* short. *Carpels* (2), each *carpel* with usually two anatropous ovules. *Fruit* a capsule, berry or drupe.

86   *Gentianaceae*

Subfamily *Oleoideae. Ovules* pendulous from apex of loculi.

Tribe 1. *Fraxineae. Petals* free or united only at base, sometimes absent. *Fruit* winged.

Only British genus (Fruit-wing distally produced):
Fraxinus.

*Tribe 2. *Syringeae. Corolla* sympetalous. *Fruit* a loculicidal capsule with winged *seeds.*
*Syringa.

Tribe 3. *Oleeae. Petals* 4, free or united. *Fruit* a berry or drupe, mostly 1-seeded.

Only British genus: Ligustrum.

Suborder 2. GENTIANINEAE. *Stamens* as many as petals. *Ovary* superior, 1–2-loc with ∞ ovules which have one integument on parietal or axile *placentae.*

Fam. 2. GENTIANACEAE. Annual or perennial, usually glabrous, bitter herbs, rarely shrubs. *Leaves* opposite, entire, exstipulate. *Flowers* cymose, usually 4–5-merous, sympetalous, hermaphrodite, actinomorphic. *Sepals* free or united. *Corolla* mostly twisted in bud. *Stamens* as many as petals, inserted on the corolla. *Carpels* (2) with usually ∞ ovules. *Ovary* mostly 1-loc with parietal placentae. *Fruit* a 2-valved capsule.

Subfamily 1. *Gentianoideae. Leaves* opposite. *Corolla* convolute or imbricate in bud. *Pollen grains* not compressed.

Tribe 1. *Chironieae. Style* slender. *Placentae* projecting.

Leaves connate. Corolla large, rotate. Stamens 6–8.   Blackstonia (= Chlora auct.).

Corolla minute, funnel-shaped. Stamens 4. Stigma peltate. **Microcala.**

Corolla salver-shaped. Stamens 4. Stigmas with two lamellae. **Cicendia.**

Corolla funnel-shaped, persistent. Stamens 5. Anthers twisted. **Centaurium.**

Tribe 2. *Swertieae. Style* short. *Placentae* not projecting.

Only British genus (Corolla subclavate): **Gentiana.**

Subfamily 2. *Menyanthoideae. Leaves* alternate. *Corolla* induplicate valvate. *Pollen grains* compressed.

Leaves 3-foliate. Corolla funnel-shaped. **Menyanthes.**

Leaves orbicular, floating. Corolla rotate. **Villarsia.**

Fam. 3. APOCYNACEAE. Shrubs or trees, rarely entirely herbaceous, often climbing. *Latex* present. *Leaves* opposite, quite entire. *Corolla* usually twisted in bud. *Stamens 4–5, anthers* basifixed. *Carpels* usually 2, free below. *Styles* united, ending in discoid or globose head which bears the stigmas. *Fruit* various.

Subfamily *Plumerioideae. Stamens* quite free or loosely attached to stigmatiferous head. *Thecae* mostly without *appendages. Seeds* usually without tuft of hairs.

Tribe *Plumerieae. Carpels* 2, free below the style.

Only British genus (disc of 2 scales, filaments clavate): **Vinca.**

C. **Sympetaly constant. Only one whorl of stamens present. Carpels usually 2, always perfectly united. Zygomorphy frequent.** (For B see page 85.)

a. Perianth hypogynous. (For *b* see page 98.)

Order 5. **Tubiflorae.** Mostly herbs. *Flowers* typically with four isomerous whorls, but the *gynaecium* usually shows reduction. When zygomorphic the *androecium* too shows reduction. *Stamens* epipetalous. *Ovules* with one *integument*.

Suborder 1. CONVOLVULINEAE. *Leaves* mostly spiral. *Flowers* usually actinomorphic. *Carpels* with few, often only two, ovules, *micropile* facing downwards. *Fruit* seldom separating into *nutlets* (never in British genera).

Fam. 1. CONVOLVULACEAE. Often left-handed twiners. *Flowers* often conspicuous, 5–4-merous. *Corolla* mostly induplicate valvate in bud. *Stamens* inserted at base of corolla-tube. *Carpels* (2), each with two basal erect ovules. *Styles* separate or united. *Fruit* usually capsular.

Subfamily 1. *Convolvuloideae*. Autotrophic, with green *foliage leaves*. *Corolla* without scales.

Tribe *Convolvuleae*. *Style* 1. *Capsule* dehiscent, 4-seeded.

Bracteoles small. Stigma slender.   **Convolvulus.**

Bracteoles large, enclosing calyx. Stigma broad.

Calystegia.

Subfamily 2. *Cuscutoideae*. Leafless parasites. *Corolla* usually with ring of *scales* beneath the *stamens*.

Only genus :   **Cuscuta.**

Fam. 2. POLEMONIACEAE. Herbs, rarely shrubs. *Leaves* spiral, exstipulate, sometimes pinnate. *Flowers* 5-merous, usually actinomorphic. *Corolla* usually twisted to right in bud. *Carpels* (3), each with usually ∞ *ovules*. *Style* 1, trifid. *Capsule* mostly loculicidal.

ERRATUM

p. 89 line 11. *For* mostly *read* rarely.

Subfamily *Polemonioideae.* Herbs. *Embryo* green, with ovate or linear cotyledons.

Tribe *Polemonieae.* *Flowers* actinomorphic. Only British genus (Leaves pinnate. Corolla almost rotate): **Polemonium.**

Suborder 2. BORRAGININEAE. As *Convolvulineae,* but *micropile* facing upwards. *Fruit* often separating into *nutlets.*

Fam. 3. BORRAGINACEAE. Mostly hispid herbs. *Leaves* alternate, undivided. *Flowers* in cincinni, 5-merous, mostly zygomorphic. *Corolla* often with *hollow folds* opposite the lobes. *Stamens* with short *filaments* and often subulate *anthers.* *Carpels* (2), each with two anatropous *ovules.* *Ovary* usually 4-loc by false septa, 4-lobed. *Style* 1, arising from between the lobes. *Stigma* simple or bifid. *Fruit* usually dividing into four nutlets.

Subfamily *Borraginoideae.* *Ovary* deeply lobed. *Fruit* of four or fewer 1-seeded nutlets. *Endosperm* absent.

Tribe 1. *Cynoglosseae.* *Nutlets* inserted by broad ventral surfaces on to conical axis, their tips not projecting beyond point of attachment.
Corolla rotate. Nutlets glabrous with inrolled borders. ***Omphalodes.**
Corolla funnel-shaped. Nutlets with hooked bristles. **Cynoglossum.**

*Tribe 2. *Eritrichieae.* Tips of *nutlets* projecting beyond narrow point of insertion. Axis conical.

Calyx with alternating teeth. Nutlets tubercled.
　　　　　　　　　　　　　　　　　　　　*Asperugo.
Calyx without alternating teeth. Nutlets with bristly
rim.　　　　　　　　　　　　　　　　　　***Lappula.**

Tribe 3. *Anchuseae. Axis* nearly flat. *Nutlets* with
concave surface of insertion which is often surrounded
by a ring.
　　a. Corolla with hollow scales.
Corolla tubular, 5-toothed, hollow scales linear, anthers
included, without appendages.　　　　**Symphytum.**
Corolla rotate, hollow scales short. Filaments with
dorsal appendages. Anthers exserted, conniving to form
a cone.　　　　　　　　　　　　　　　**Borrago.**
Corolla salver-shaped, hollow scales short. Stamens
included, with no appendages.
　　　　　　　　Anchusa (including Lycopsis).
　　b. Corolla without hollow scales.　**Pulmonaria.**

Tribe 4. *Lithospermeae. Nutlets* erect with small
flat surface of insertion.
　　No calyx-tube. Corolla imbricate in bud. Stamens
included. Nutlets stony.　　　　　　**Lithospermum.**
Calyx-tube short. Corolla imbricate in bud. Stamens protruding. Nutlets soft.　　　　**Mertensia.**
Calyx-tube long. Corolla twisted in bud. Nutlets
smooth.　　　　　　　　　　　　　　**Myosotis.**

Tribe 5. *Echieae. Flowers* zygomorphic. *Corolla*
usually without scales.
　　Only genus (Calyx regular):　　　　**Echium.**

Suborder 3. VERBENINEAE. *Leaves* mostly opposite
or whorled. *Flowers* mostly zygomorphic. Each *carpel*

with two ovules, rarely with only one. *Fruit* often of nutlets.

Fam. 4. VERBENACEAE. Herbs, shrubs or trees. *Leaves* mostly opposite. *Flowers* mostly zygomorphic. *Calyx* tubular. *Corolla* often with curved *tube* and 2-lipped *limb.* *Stamens* mostly 4, didynamous. *Carpels* mostly (2), each with two *ovules*; *micropyle* facing downwards. *Ovary* usually ultimately 4-loc. *Style* terminal, simple. *Fruit* mostly drupaceous (nutlets in British species).

Tribe *Verbeneae.* Mostly herbs. *Flowers* in spikes or racemes. *Loculi* of ovary divided or not. No *endosperm.*

Only British genus: **Verbena.**

Fam. 5. LABIATAE. Aromatic herbs and shrubs. *Leaves* and *branches* opposite or whorled. *Flowers* usually zygomorphic, in cymes often forming *false whorls.* K 5. C 5, *limb* often 2-lipped. A 4, didynamous, or two *stamens* and two *staminodes,* or two *stamens* and no *staminodes,* 5th *stamen* rarely represented by *staminode.* G (2), each with two erect anatropous *ovules.* The *carpels* are folded in between the *ovules.* *Fruit* dividing into four *nutlets,* or fewer by abortion.

Subfamily 1. *Ajugoideae. Calyx-tube* 10–11-ribbed. *Corolla* 1 or 2-lipped, rarely almost actinomorphic. *Stamens* 4, rarely 2. *Ovary* usually shortly divided, or to one-third its length. *Nutlets* obovoid with dry *pericarp. Endosperm* very scanty. *Embryo* straight.

Tribe 1. *Ajugeae. Calyx* 10-ribbed. *Corolla* either almost actinomorphic or (as in British genera) with very

short *upper lip* and *lower lip* reflexed. *Stamens* 4 or 2.
Anthers 2-celled. *Nutlets* warty or reticulate.
Corolla persistent, with ring of hairs within. Upper
lip 2-lobed.   **Ajuga.**
   Corolla deciduous, without ring of hairs. Upper lip
absent.   **Teucrium.**

   Subfamily 2. *Scutellarioideae.* *Calyx* 2-lipped.
Corolla 2-lipped, with helmet-shaped upper lip. *Stamens* 4.
Ovary 4-lobed. *Nutlets* more or less globular, with dry
pericarp, attached by small basal insertion to gynophore-
like torus. *Seed* transverse. *Endosperm* absent. *Embryo*
bent.
   (Calyx-tube dilated above to form hollow pouch.)
                                                Scutellaria.

   Subfamily 3. *Stachydoideae.* *Calyx-tube* with
5–15 ribs. *Corolla* almost actinomorphic or markedly
zygomorphic. *Stamens* 4 or 2. *Ovary* 4-lobed. *Nutlets*
ovoid, obovoid or tetrahedric, with dry pericarp and
small basal attachment. *Seed* straight. *Embryo*
straight.

   Tribe 2. *Marrubieae.* *Calyx* tubular or campanulate,
5–10-toothed. *Corolla* usually hidden by *calyx.* *Stamens*
and *style* included.
   Only British genus: **Marrubium.**

   Tribe 3. *Nepeteae.* *Calyx* 15-ribbed. *Corolla* 2-lipped.
Stamens 4, posterior (inner) pair longer or alone present.
   Only British genus: **Nepeta.**

   Tribe 4. *Stachyeae.* *Calyx* 5–10-ribbed. *Upper lip*
of corolla concave. *Stamens* 4, parallel, ascending under
the upper lip, *anterior* (*outer*) pair longer.

Subtribe 1. *Brunellinae.* *Calyx* 2-lipped. Lips closing together in fruit. *Upper lip of corolla* concave. Only British genus : **Brunella.**

Subtribe 2. *Melittinae.* Calyx broadly campanulate not closing in fruit. *Corolla-tube* dilated throughout. Only British genus : **Melittis.**

Subtribe 3. *Lamiinae.* *Calyx* tubular not closing in fruit. *Corolla-tube* not dilated below.

A. Anther-thecae transverse during dehiscence. Corolla with hollow scales on palate. **Galeopsis.**

B. Dehiscing anther-thecae vertical or oblique.

1. Nutlets sharply 3-angular with obtuse apex. Leaves serrate. Calyx teeth not spinous. Throat of corolla dilated. **Lamium.**

Leaves incised. Calyx teeth spinous. Throat of corolla not dilated. ***Leonurus.**

2. Nutlets ovoid with rounded apex. Calyx funnel-shaped, teeth dilated at base or united to form a ring. **Ballota.**

Calyx tubular or campanulate, teeth 3-angular, not dilated or united. **Stachys.**

Tribe 5. *Salvieae.* *Calyx* campanulate or tubular. *Corolla* with helmet-shaped or sickle-shaped upper lip. *Stamens* 2 (the anterior pair). *Connective* linear, bearing one fertile theca. Only British genus : **Salvia.**

Tribe 6. *Satureieae.* *Calyx* with five equal teeth or 2-lipped. *Corolla* with flat lobes, almost actinomorphic or 2-lipped. *Stamens* 4 or 2, equal or the anterior pair longer.

1. Corolla distinctly 2-lipped. Stamens 4, ascending under upper lip, spreading above.
Corolla-tube recurved. Upper lip concave. **Melissa.**
Corolla-tube almost straight. Upper lip flat.
Satureia (Calamintha auct.).
2. Corolla distinctly 2-lipped. Stamens 4, spreading.
Erect. Leaves broad. Calyx tubular. **Origanum.**
Procumbent. Leaves small. Calyx 2-lipped.
Thymus.
3. Corolla almost actinomorphic. Stamens spreading.
Fertile stamens 4. **Mentha.**
Two anterior stamens alone fertile. **Lycopus.**

Suborder 4. SOLANINEAE. *Flowers* actinomorphic, or more often zygomorphic, typically 5-merous. *Stamens* 5 or 4 or 2. *Carpels* rarely (5), mostly (2), usually with ∞ *ovules*. *Fruit* mostly capsular, never dehiscent quite to the base.

a. Vascular bundles bicollateral. (For β see p. 95.)

Fam. 6. SOLANACEAE. Herbs or shrubs with spiral *leaves*. *Flowers* terminal, single and often supra-axillary, or in cymose, often supra-axillary, *inflorescences*, mostly 5-merous, hermaphrodite, actinomorphic, or rarely zygomorphic. *Corolla-lobes* mostly folded in bud. *Stamens* 5 (in zygomorphic flowers one may be staminodal). *Carpels* (2), oblique. *Ovules* ∞ on the septum. *Style* 1. *Fruit* a berry or capsule. *Seeds* with endosperm.

Tribe 1. *Solaneae. Ovary* 2-locular.
Shrubs. Corolla salver-shaped. Stamens attached above middle of corolla-tube. Fruit a berry. *Lycium.
Herbs. Corolla campanulate. Stamens attached at base of tube. Fruit a berry. **Atropa.**

Herbs. Corolla funnel-shaped. Capsule opening by lid. **Hyoscyamus.**

Herbs or shrubs. Corolla rotate. Anthers conniving. Fruit a berry. **Solanum.**

*Tribe 2. *Datureae.* Ovary 4-locular by growth of *secondary dissepiments.*

(Capsule with spines.) *Datura.

β. Bundles collateral. (For α see p. 94.)

I. *Ovary* 2-locular, usually with ∞ axile *ovules.* (For II see p. 97.)

Fam. 7. SCROPHULARIACEAE. Herbs, shrubs or trees. *Leaves* spiral or opposite. *Inflorescence* various, *flowers* never terminal. *Flowers* 5-merous, hermaphrodite, more or less zygomorphic. *Calyx* usually persistent. *Stamens* rarely 5, mostly 4 or 2. G (2), median (not oblique), each with ∞–few anatropous *ovules* on the *dissepiment. Style* 1. *Fruit* a capsule or berry.

Subfamily 1. *Pseudosolanoideae. Leaves* mostly spiral. The two posterior (upper) *corolla-lobes,* or the *upper lip,* cover the lateral lobes in bud. Five fertile *stamens* present.

Tribe 1. *Verbasceae. Corolla-tube* short or absent. Only British genus (Corolla rotate. Stamens 5): **Verbascum.**

Subfamily 2. *Antirrhinoideae. Lower leaves* at least opposite. *Aestivation* as in *Pseudosolanoideae. Fifth (posterior) stamen* staminodal or absent.

Tribe 2. *Antirrhineae. Flowers* axillary or racemose, zygomorphic. *Corolla-tube* distinct, spurred or saccate at base. *Capsule* usually opening by pores.

1. Corolla-tube saccate at base.   **Antirrhinum.**

2. Corolla-tube spurred at base. Leaves narrow, pinnately veined. Flowers in terminal racemes. Capsule with valves.   **Linaria.**

Leaves narrow, pinnately veined. Flowers solitary. Capsule with pores.   **Elatinoides.**

Leaves cordate, digitately nerved and lobed. Capsule with small trifid valves.   ***Cymbalaria.**

Tribe 3. *Cheloneae. Flowers* ultimately cymose, zygomorphic. *Corolla-tube* not spurred or saccate. Only British genus: **Scrophularia.**

Tribe 4. *Gratioleae. Inflorescence* racemose or *flowers* axillary. *Corolla-tube* not spurred or saccate.

Leaves opposite. Flowers large. Stigma 2-lamellate.   ***Mimulus.**

Leaves fascicled. Flowers minute. Stigma clavate.   **Limosella.**

Subfamily 3. *Rhinanthoideae.* The two posterior (upper) *corolla-lobes* or the *upper lip* are covered in bud by the lateral *corolla-lobes.*

a. Corolla-lobes all nearly flat.

Tribe 5. *Digitaleae.* Not parasitic. *Anther-cells* united at end.

Corolla-tube short. Stamens 4–8, equal.   **Sibthorpia.**

Corolla-tube short. Stamens 2, equal.   **Veronica.**

Large herbs. Corolla-tube long, dilated. Stamens 4, didynamous.   **Digitalis.**

Small herbs. Corolla-tube cylindric. Stamens 4, didynamous.   ***Erinus.**

Orobanchaceae, Lentibulariaceae 97

b. The two upper *corolla-lobes* form a helmet-shaped upper lip.

Tribe 6. *Rhinantheae.* Partial and entire parasites.

1. Leaves opposite, narrow. Seeds 1–4, large, not winged. **Melampyrum.**

2. Leaves opposite. Seeds ∞, minute, not winged. Upper lip of corolla almost entire. **Bartschia.** Upper lip of corolla 2-lobed. **Euphrasia.**

3. Leaves opposite. Seeds few, winged. **Rhinanthus.**

4. Leaves alternate. Seeds ∞, minute, not winged. **Pedicularis.**

II. *Ovary* 1-locular with ∞ *ovules* on more or less parietal *placentae.* (For I see p. 95, for III see below.)

Fam. 8. OROBANCHACEAE. Annual and perennial, usually brownish, parasitic herbs. *Leaves* scale-like. *Flowers* usually racemose, 5-merous, zygomorphic. *Corolla* distinctly 2-lipped, tube curved. *Stamens* 4, didynamous. *Carpels* (2), median. Each *carpel* with two parietal placentae with ∞ ovules. *Style* 1. *Fruit* a loculicidal capsule.

Subterranean leaves fleshy and with cavities. Flowers racemed. Stamens shortly exserted. **Lathraea.**

Subterranean leaves membranous or 0. Flowers sub-spikate. Stamens included. **Orobanche.**

III. *Ovary* usually 1-celled with free central *placenta* bearing ∞ *ovules.*

Fam. 9. LENTIBULARIACEAE. Herbs, mostly aquatic, or growing in wet places. *Roots* often absent. *Leaves*

c. 7

radical or whorled, undivided or multifid. *Flowers* 5-merous, mostly zygomorphic. *Corolla* usually distinctly 2-lipped. *Stamens* usually 2. *Carpels* (2), median. *Ovules* many on the free central *placenta. Fruit* capsular.

Tribe *Utricularieae. Corolla* zygomorphic with distinct *spur.*
Land plants. Leaves radical, entire. Corolla throat open.   **Pinguicula.**
Aquatic plants. Leaves floating, multifid. Corolla throat closed.   **Utricularia.**

Order 6. **Plantaginales.** Mostly herbs. *Leaves* usually spiral. *Flowers* actinomorphic, 4-merous (except *gynaecium*). *Corolla* scarious. One whorl of *stamens. Carpels* usually 2.

Fam. PLANTAGINACEAE. Scapigerous herbs. *Leaves* mostly radical, parallel-veined. *Flowers* small, anemophilous, in spikes. K 4, persistent. C 4, scarious, *lobes* spreading. A 4, inserted on corolla-tube (hypogynous in *Litorella*). *Anthers* large, versatile. *Carpels* usually (2). *Ovary* 2-loc or 4-loc by spurious septum. *Ovules* usually several, anatropous, peltate, on the septum. *Fruit* a capsule dehiscing transversely or a nut. *Seeds* with endosperm.
Land plants. Flowers many, spiked, hermaphrodite. Capsule dehiscing transversely.   **Plantago.**
Water plant. Flowers few, monoecious. Fruit a nut.   **Litorella.**

b. Perianth epigynous. (For *a* see p. 87.)
a. Stamens free. (For β see page 102.)

Order 7. **Rubiales.** Herbs, shrubs and trees. *Leaves* opposite, usually undivided. *Flowers* 5–4-merous, isomerous or *androecium* and *gynaecium* oligomerous, usually actinomorphic. *Ovary* inferior with one or several loculi, each with ∞–1 anatropous *ovules.*

A. *Stamens* as many as corolla-lobes. (For B see page 101.)

Fam. 1. RUBIACEAE. Herbs, shrubs or trees, with opposite entire *leaves.* *Stipules* always present, interpetiolar (foliaceous in British species). *Flowers* 5–4-merous, usually actinomorphic. *Sepals* mostly valvate. *Carpels* mostly (2), more rarely (1–∞), each carpel with 1–∞ anatropous *ovules.* *Style* 1, with capitate or divided *stigma.* *Fruit* various.

Subfamily *Coffeoideae.* *Carpels* each with one *ovule.*

Tribe *Galieae.* Herbs. *Leaves* and *foliaceous stipules* forming whorls. *Ovules* attached to septum.

A. Sepals distinct, more than half as long as ovary. Corolla funnel-shaped with long tube. **Sherardia.**

B. Sepals minute or absent.

a. Corolla funnel-shaped. **Asperula.**

b. Corolla rotate.

Flowers often 5-merous. Fruit succulent. **Rubia.**

Flowers 4-merous. Fruit dry. **Galium.**

Fam. 2. CAPRIFOLIACEAE. Usually shrubs or small trees. *Leaves* opposite, usually exstipulate. *Stipules* when present (as in *Sambucus*) not interpetiolar. *Flowers* actinomorphic or zygomorphic. *Carpels* (2–5), each with

7—2

$1-\infty$ axile, pendulous *ovules*. *Style* simple or divided. *Fruit* a berry or drupe, rarely a capsule.

Tribe 1. *Sambuceae*. *Leaves* pinnately divided. *Anthers* opening outwards. *Style* very short. *Fruit* a drupe. (Very near *Valerianaceae*.)
Only British genus: **Sambucus.**

Tribe 2. *Viburneae*. *Leaves* entire or lobed. *Anthers* opening inwards. *Carpels* (1–5), each with one *ovule*. *Style* usually very short. *Fruit* a berry or drupe.
Only British genus: **Viburnum.**

Tribe 3. *Linnaeeae*. *Leaves* undivided. *Anthers* opening inwards. *Carpels* (3–4), of which two have ∞ sterile *ovules* and 1–2 have one fertile *ovule*. *Style* long.
Stems slender, creeping. Peduncles 2-flowered. Ovary 3-loc. **Linnaea.**
Shrubs. Flowers in racemes. Ovary 4-loc.
*Symphoricarpus.

Tribe 4. *Lonicereae*. *Leaves* usually undivided. *Anthers* introrse. *Carpels* (2–5–8) all with ∞ ovules. *Style* usually very long.
Only British genus (Flowers zygomorphic):
Lonicera.

Fam. 3. ADOXACEAE. A small succulent herb. *Leaves* ternate. *Flowers* in 5-flowered heads, terminal 4, lateral 5-merous, homochlamydeous. *Calyx* (*bracts and bracteoles*) 2–3-lobed. *Stamens* 4 or 5–6, split to the base. *Carpels* (3–5), each with one pendulous ovule. *Fruit* a drupe with 1–3 *stones*. The position of this monotypic genus is very uncertain.
Only genus: **Adoxa.**

B. *Stamens* fewer than *corolla-lobes.* *Ovary* inferior
always with only one fertile *loculus* with one pendulous
ovule. (For A see page 99.)

Fam. 4. VALERIANACEAE. Herbs, rarely shrubs,
with opposite exstipulate *leaves.* *Inflorescence* cymose.
Flowers rather irregular. *Calyx* usually an involute
pappus expanding as fruit ripens. *Corolla* (5) or (3–4),
often with pouch or spur. *Stamens* 1–4. *Anthers* exserted, versatile. *Carpels* (3), of which one has one
pendulous anatropous ovule, the other two empty.
Style 1. *Stigmas* 1–3.

Tribe *Valerianeae.* *Stamens* usually 3, more rarely
2 or 1.
Calyx obscurely 5-toothed. Stamens 3.
Valerianella.
Calyx forming pappus. Corolla-tube pouched at
base. Stamens 3. **Valeriana.**
Calyx forming pappus. Corolla-tube spurred at
base. Stamen 1. ***Kentranthus.**

Fam. 5. DIPSACACEAE. Usually herbs. *Leaves*
opposite, exstipulate. *Flowers* usually in heads. *Outer
bracts* often forming involucre, *inner* axillant to flowers
or absent. *Bracteoles* forming *epicalyx* (*involucel*).
Flowers hermaphrodite, mostly zygomorphic. *Stamens*
4 or fewer. *Carpels* (2), but *ovary* 1-loc with one pendulous *ovule.* *Style* 1. *Stigmas* 1–2. *Fruit* indehiscent,
enclosed in *epicalyx.*
Involucral-bracts very long, rigid. Floral-bracts
spinous, longer than flowers. **Dipsacus.**
Involucral-bracts short. Floral-bracts short or absent. **Scabiosa.**

β. *Stamens conniving or partly united.* (For a see p. 98.)
Order 8. **Cucurbitales.** Mostly climbing herbs.
Flowers typically 5-merous. *Stamens* 5, rarely all free,
usually appearing as 3 by union of two pairs, or all 5
may be united into a *synandrium.*

Fam. CUCURBITACEAE. Usually annual herbs of
rapid growth climbing by means of *tendrils. Leaves*
spiral, mostly lobed. *Flowers* cymose, usually diclinous,
actinomorphic. *Flower-axis* cup-shaped. *Stamens* 5 free
(rare), or more often apparently 3 by fusion. *Anthers*
often twisted. *Carpels* usually (3), each with two axile
revolute *placentae* bearing usually ∞ anatropous *ovules.*
Ovary inferior, usually 3-locular. *Style* 1. *Stigmas*
usually 3. *Fruit* usually a berry. No *endosperm.*

Tribe *Cucurbiteae. Stamens* 5, mostly two pairs
united and one free. *Anthers* with ∞ -shaped or U-shaped
pollen-sacs.

Only British genus :   Bryonia.

Order 9. **Campanulatae.** Mostly herbs. *Flowers*
typically 5-merous with isomerous *androecium* and
oligomerous *gynaecium. Anthers* often united. *Ovary*
inferior, with several *loculi* each with ∞ –1 *ovules,* or 1-loc
with one *ovule.*

Fam. 1. CAMPANULACEAE. Usually herbs with
latex. *Leaves* spiral, exstipulate. *Flowers* often large,
usually 5-merous, hermaphrodite, and actinomorphic.
Stamens free or united. *Anthers* introrse. *Carpels*
mostly (2–5) with ∞ anatropous *ovules. Style* 1, often
with device for brushing up and collecting *pollen. Ovary*
mostly inferior, of several loculi, rarely 1-loc. *Placentae*
usually axile. *Fruit* usually capsular. *Seeds* with
fleshy endosperm.

Subfamily 1. *Campanuloideae*. *Flowers* actino-morphic, rarely slightly zygomorphic. *Anthers* usually free.

Tribe *Campanuleae*. *Flowers* symmetrical. *Corolla-lobes* valvate in bud.

Subtribe 1. *Campanulinae*. Ovary always quite inferior. Fruit usually a capsule dehiscing laterally.
Flowers often capitate. Corolla-lobes linear.

Phyteuma.

Flowers racemose. Corolla-lobes broad. Capsule short. **Campanula.**

As *Campanula*, but ovary and capsule slender.

Specularia.

Subtribe 2. *Wahlenbergiinae*. Ovary inferior or almost superior. Fruit usually a capsule dehiscing apically.
Flowers capitate. Corolla-lobes almost free. Anthers united at base. **Jasione.**

Inflorescence lax. Corolla campanulate. Anthers free. **Wahlenbergia.**

Subfamily 2. *Lobelioideae*. Flowers zygomorphic and resupinate. Anthers united.
Only British genus : **Lobelia.**

Fam. 2. COMPOSITAE. Mostly herbs. *Leaves* spiral or opposite, exstipulate, sometimes with auricles. *Flowers* in involucrate *heads*. *Involucral-bracts* spiral or whorled, sometimes united, usually not bearing flowers in their axils. *Floral-bracts* chaffy, setaceous, petaloid, or absent. *Flowers* 5-merous, hermaphrodite, or ♂ and ♀ flowers separate, or sterile, actinomorphic or zygomorphic. *Calyx* rarely distinctly developed, usually represented by hairs, bristles or scales forming the *pappus*. *Corolla* tubular, ligulate, or 2-lipped (in Tribe *Mutisieae*, not represented in Britain). *Stamens* inserted on the *corolla-tube*, *filaments* usually free. *Anthers* usually united, forming a tube, dehiscing inwards. *Carpels* (2), median, but the *ovary* is 1-locular with one ascending, anatropous *ovule*. *Style* bifid in the fertile flowers, the two arms stigmatic on their inner surfaces. *Fruit* a 1-seeded, inferior nut. *Seeds* without endosperm.

Subfamily 1. *Tubuliflorae*. No *latex* present. *Corolla* of *disc-flowers* never ligulate.

Tribe 1. *Eupatorieae*. *Leaves* usually opposite. *Flowers* all hermaphrodite and alike. *Corolla* tubular, never pure yellow. *Anthers* basifixed, obtuse at base. *Style-arms* hairy outside, but the style has no ring of hair below the bifurcation.

Only British genus: **Eupatorium.**

Tribe 2. *Astereae*. *Leaves* spiral. *Heads* mostly with ♀ ligulate *ray-flowers* and ☿ tubular *disc-flowers*. *Anthers* as in *Eupatorieae*. *Style* without ring of hairs below bifurcation, arms thickly hairy above.

Ray-flowers 1-seriate, yellow: disc-flowers yellow. Pappus scabrid. **Solidago.**

Ray-flowers 1-seriate, white or pink: disc-flowers yellow. Pappus absent. **Bellis.**

Ray-flowers 1-seriate, white, blue or purple: disc-flowers yellow. Pappus many-seriate, persistent. **Aster.**

As *Aster*, but ray-flowers 2 or more seriate.

Erigeron.

Compositae 105

Tribe 3. *Inuleae. Leaves* spiral. *Ray-flowers* mostly ♀ : *disc-flowers* often 4-merous. *Anthers* with tails at the base. *Pappus* mostly of hairs.

1. Corolla of ♀ (ray) flowers filiform. Chaffy floral-bracts present. **Filago.**

2. As 1 but floral-bracts absent. Heads dioecious.

Antennaria (including ***Anaphalis**).

Heads with ♂ and ♀ flowers. **Gnaphalium.**

3. Corolla of ♀ (ray) flowers ligulate. Floral-bracts absent.

Pappus of scabrid hairs. **Inula.**

Pappus of an inner row of scabrid hairs and an outer row of scales. **Pulicaria.**

Tribe 4. *Heliantheae. Leaves* usually opposite. *Involucral-bracts* without broad scarious margin. *Disc-flowers* actinomorphic. *Floral-bracts* present. *Anthers* basifixed with obtuse base. *Style* with hairs mostly above bifurcation. *Pappus* never of hairs.

Subtribe 1. *Ambrosiinae.* Corolla of ♀ flowers reduced or absent. Anthers sometimes quite free.

Xanthium.

Subtribe 2. *Coreopsidinae.* All flowers with perfect corollas. Corolla of ray-flowers early deciduous. Pappus absent or of bristles. **Bidens.**

Subtribe 3. *Galinsoginae.* As above, but pappus of ciliate scales. ***Galinsoga.**

Tribe 5. *Anthemideae.* As *Heliantheae* but often aromatic, *leaves* spiral, and *involucral-bracts* with broad scarious margins. *Pappus* minute or absent.

Subtribe 1. *Anthemidinae.* Floral-bracts present.
Ray-flowers with oblong ligules. Fruit not much
compressed.   **Anthemis.**
Ray-flowers with short, broad ligule. Fruit much
compressed.   **Achillea.**
Ray-flowers absent. Corolla compressed, with corky
appendages.   **Diotis.**
Subtribe 2. *Chrysantheminae.* Floral-bracts absent.
Ray-flowers ligulate. Involucral-bracts 1–2-seriate,
all equal.   **Matricaria.**
Ray-flowers ligulate. Involucral-bracts many-seriate,
outer shorter.   **Chrysanthemum.**
Flowers all tubular, 4-merous.   ***Cotula.**
Flowers all tubular, 5-merous. Axis of head broad.
                                               Tanacetum.
Flowers all tubular, 5-merous. Axis of head narrow.
                                               Artemisia.

Tribe 6. *Senecioneae. Leaves* spiral. *Involucral-
bracts* usually 1-seriate, without scarious margin. *Flowers*
usually all yellow. *Floral-bracts* absent. *Anthers* usually
obtuse at base. *Style-arms* various. *Pappus* usually of
copious, soft hairs.

* Leaves often very large, produced after the flowers.
Style-arms of disc-flowers connate.
Heads in racemes, purple or white, outer flowers
tubular.   **Petasites.**
Heads solitary, yellow, outer flowers ligulate.
                                               Tussilago.

** Leaves radical and cauline. Style-arms of disc-
flowers separate, truncate.
Perennial herbs. Involucral-bracts in several series
all equal.   ***Doronicum.**

Annual and perennial herbs. Involucral-bracts 1-seriate or with a few smaller ones at the base.

                                                           Senecio.

*Tribe 7. *Calenduleae. Leaves* spiral. *Axis of head* naked. *Disc-flowers* hermaphrodite, functionally ♂ (not ripening fruit). *Ray-flowers* ♀. *Anthers* pointed at base. *Pappus* absent.
(Fruit bent inwards.)   ***Calendula.**

Tribe 8. *Cynareae. Leaves* spiral, often spinous. *Involucral-bracts* many-seriate, often spinous. *Axis of head* usually with *bristles* which have no definite relationship to the flowers. *Flowers* often purple, usually all tubular, all hermaphrodite, or outer ♀ or neuter. *Anthers* usually with tails at base. *Style* usually with ring of hairs or swelling.

Subtribe 1. *Echinopsinae.* Heads 1-flowered, themselves arranged in heads of the second order. Flowers blue or white.   ***Echinops.**

Subtribe 2. *Carlininae.* Heads many-flowered. Fruit with straight insertion, usually covered with silky hairs. Pappus of scales or of 1-seriate, feathery hairs.
Only British genus :   **Carlina.**

Subtribe 3. *Carduinae.* Heads many-flowered. Fruit with straight insertion, usually glabrous. Pappus usually of many-seriate hairs.

A. Pappus hairs simple.
Outer involucral-bracts hooked.   **Arctium.**
Outer involucral-bracts not hooked.
Filaments free, warted or hairy. Pappus hairs connate at base.   **Carduus.**

Filaments connate into a sheath. Pappus free.
Silybum.

B. Pappus hairs all, or inner ones only, feathery.
Involucral-bracts without spines. Filaments glabrous.
Saussurea.
Involucral-bracts with spines. Filaments warty or
hairy. Fruit compressed or terete. **Cnicus.**
Involucral-bracts with spines. Filaments nearly
glabrous. Fruit 4-angled. **Onopordon.**

Subtribe 4. *Centaureinae.* Heads many-flowered.
Fruit with oblique insertion, glabrous or hairy. Pappus
many-seriate.
Involucral-bracts without appendages. All flowers
hermaphrodite. **Serratula.**
Involucral-bracts with scarious or thorny appendages.
Outer flowers quite sterile and larger than inner ones.
Centaurea.

Subfamily 2. *Liguliflorae. Latex* present. *Corolla*
of all flowers ligulate.

Tribe 9. *Cichorieae. Leaves* spiral. *Style-arms*
cylindrical, hairy, obtuse.

Subtribe 1. *Hyoseridinae.* Fruit obtuse. Pappus of
scales. (Flowers blue in British species.) **Cichorium.**

Subtribe 2. *Lapsaninae.* Involucral-bracts almost
equal. Fruit obtuse. Pappus absent. (Flowers yellow
in British species.)
Leaves all radical. Fruit crowned with ring.
Arnoseris.
Leaves radical and cauline. Fruit without ring.
Lapsana.

Subtribe 3. *Crepidinae.* Neither stellate nor woolly hairs present. Only inner involucral-bracts equal. Fruit with simple or feathery pappus.

Pappus of feathery hairs.   **Picris.**

Pappus of simple hairs.   **Crepis.**

Subtribe 4. *Hieraciinae.* Stellate or woolly hairs often present on leaves. Fruit rounded at apex. Pappus rough, brown.   **Hieracium.**

Subtribe 5. *Hypochaeridinae.* Involucral-bracts all imbricate, the inner ones keeled after flowering. Fruit beaked. Pappus simple or feathery.

Narrow, membranous floral-bracts present. Beak of inner fruits long. Pappus hairs feathery. **Hypochaeris.**

No floral-bracts. All beaks short. Pappus hairs feathery.   **Leontodon.**

No floral-bracts. All beaks long. Pappus hairs simple.   **Taraxacum.**

Subtribe 6. *Lactucinae.* Leaves usually pinnatifid. Inner involucral-bracts equal and unaltered after flowering. Fruit acute or beaked, with numerous simple pappus hairs.

Fruit not beaked, scarcely narrowed above. **Sonchus.**

Fruit narrowed or beaked above.   **Lactuca.**

Subtribe 7. *Scorzonerinae.* Leaves linear, entire. Fruit usually with long beak. Pappus simple or feathery.

Only British genus (Involucral-bracts 1-seriate):

Tragopogon.

INDEX

C.

8

8—3

Trifolium 60
Triglochin 13
Triglochineae, tribe 13
Trigonella 60
Trinia 78
Trisetum 18
Triticum (including Agropyrum and Secale) 20
Trollius 47
Tsuga (including Pseudotsuga) 9
Tubiflorae, order 88
Tubuliflorae, subfamily 104
Tulipa 26
Tulipeae, tribe 26
Tunica 45
Tussilago 106
Typha 11
Typhaceae, family 11

Ulex 60
Ulmaceae, family 36
Ulmaria 59
Ulmarieae, tribe 59
Ulmoideae, subfamily 36
Ulmus 36
Umbellales (Umbelliflorae), order 74
Umbelliferae, family 75
Umbelliflorae, order 74
Urereae, tribe 37
Urtica 37
Urticaceae, family 37
Urticales, order 36
Utricularia 98
Utricularieae, tribe 98

Vaccinieae, tribe 82
Vaccinioideae, subfamily 82
Vaccinium (including Oxycoccos) 82

Valeriana 101
Valerianaceae, family 101
Valerianeae, tribe 101
Valerianella 101
Vallisnerioideae, subfamily 15
Verbasceae, tribe 95
Verbascum 95
Verbena 91
Verbenaceae, family 91
Verbeneae, tribe 91
Verbenineae, suborder 90
Veronica 96
Viburneae, tribe 100
Viburnum 100
Vicia 61
Vicieae, tribe 61
Villarsia 87
Vinca 87
Viola 71
Violaceae, family 71
Violeae, tribe 71
Viscaria 44
Visceae, tribe 38
Viscoideae, subfamily 38
Viscum 38

Wahlenbergia 103
Wahlenbergiinae, subtribe 103
Wolffia 23
Wolffoideae, subfamily 23
Woodsia 5
Woodsieae, tribe 5

Xanthium 105

Zannichellia 12
Zannichellieae, tribe 12
Zostera 12
Zostereae, tribe 12

Printed in the United States
By Bookmasters